Travels on the Western Waters
John Francis McDermott, General Editor

Journey Through a Part of the United States ✤ ✤ of North America in the Years 1844 to 1846 ✤

By DR. ALBERT C. KOCH

Translated and Edited by Ernst A. Stadler

Foreword by John Francis McDermott

SOUTHERN ILLINOIS UNIVERSITY PRESS

Carbondale and Edwardsville

Feffer & Simons, Inc.
London and Amsterdam

Library of Congress Cataloging in Publication Data

Koch, Albrecht Karl.
   Journey through a part of the United States of North America in the years 1844 to
1846.

   (Travels on the western waters)
   Translation of Reise durch einen Theil der Vereinigten Staaten von Nordamerika in
den Jahren 1844 bis 1846.
   Bibliography: p.
   1.   United States—Description and travel—1783–1848.
   2.   Natural history—United States.   I.   Title.
E165.K7613   917.3   76–188701
ISBN 0–8093–0581–X

For Frances

Contents

List of Illustrations

ix

Foreword

G E R M A N - B O R N Albert C. Koch at first glance seems to have been one of those oddities so often found on the American scene in the early nineteenth century. After some years of obscure residence in St. Louis he came to public attention in January 1836 when he announced in the *Missouri Republican* the opening of a "museum." There for only twenty-five cents the curious could gaze on an Egyptian mummy with sarcophagus, an Indian mummy from a cave in Kentucky, cosmoramic views of the Battle of Austerlitz, the tunnel under the Thames, the French revolution of 1830, and other such pictorial news, wax figures of the Siamese Twins, of a Chinese lady in her native dress, Jim Crow, Zip Coon and other personages, a large collection of stuffed birds and animals, and other intriguing and educational phenomena. In succeeding years Koch added portraits of Santa Ana, president of Mexico, General Cos, his brother-in-law, and Osceola, the Seminole Indian chief, and, for a short time, five live alligators. One of his rooms was often given over to shows such as the "splendid and very unique ASIATIC ENTERTAINMENTS" of "THE GREAT PERSIAN KOULAH," Master Platt the ventriloquist, the well-known Chapman Family in a "Musical OLIO," and the "FEAST OF MIRACLES" staged by "Miss Zelina-Kha-Nourhina, very advantageously known as the *Peri of the Caspian*" and her father "the Professor of Hindoo Deceology."

But all this was simply a means of earning a living. Actually, we discover in 1838, Koch was a passionate paleontologist. In October of that year, on hearing of a deposit of bones in Gasconade County, Missouri, he rushed out to disinter the remains of an animal the size of an elephant. In May 1839, twenty-two miles south of St. Louis he dug out other prehistoric remains. In the spring of 1840 he came back to St. Louis from an excursion into central Missouri with two skeletons of a super-mastodon which he named the *Missourium*. This earnest, enthusiastic, and hardworking amateur of science made some errors in assembling his fabulous animal, but the British Museum in 1843 thought enough of his discovery to pay him the considerable sum of £ 1300 for the reconstruction (and there it can be seen today, though somewhat reduced in size). To the Royal Museum of Berlin, about the same time, he disposed of the skeleton of a *Zeuglodon*, another of his "finds." Obviously, Koch's contribution to science was being recognized.

My first acquaintance with this fascinating man began about thirty-five years ago when I was gathering material for an account of museums in early St. Louis. Since Koch's activities were only a part of the survey I had undertaken, I was content to rest on the details I could extract from the local press, from Koch's own pamphlets in English, and from a few reports by visitors to the city. In due time I read my paper before the Missouri Academy of Science and then in my impetuous way rushed it into print in the *Missouri Historical Society Bulletin* in 1948.

But I never forgot Dr. Koch and his astounding fossils, and I was happy recently to interest Ernst A. Stadler, Bavarian-born St. Louisan, in translating for the "Travels on the Western Waters series" the diary kept by this devoted enthusiast on his second tour in America after he had disposed of his *Missourium* and other bones in Europe. Leaving his family in Germany, he returned to the United States in 1844 and from Martha's Vineyard he roamed westward, down the Ohio and up and down the Mississippi and the Alabama rivers, following up every rumor about deposits of bones, a tour that reached its climax in his excavation of the *Hydrarchos*, a serpentlike monster equal in magnificence to his *Missourium*.

This diary, published by Koch in Dresden in 1847, has become very rare. Mr. Stadler's able translation of a difficult text and his pursuit of his author's life in America will bring to this odd paleontologist some of the attention he should have for his untiring efforts in gathering prehistoric remains and for the contribution he made toward establishing the great antiquity of man in the western hemisphere. Though Koch's interests differed greatly from those of Timothy Flint and Francis Baily, whose narratives have already been published in this series, his diary, too, is representative of travel on the western waters.

Southern Illinois University   *John Francis McDermott*
Edwardsville   General Editor

Preface

THE GERMAN edition of excerpts from the travel diary of Albert C. Koch was published under the title *Reise durch einen Theil der Vereinigten Staaten von Nordamerika in den Jahren 1844 bis 1846* in Dresden in 1847, and was intended for the German reader of that time, who was avid to hear all about America. Koch, the pilgrim with a pickax, was a geologist and naturalist who exhumed remains of prehistoric animals and collected petrifactions in many sections of North America. This wanderer between two worlds delighted in amazing the inhabitants of the Old World with the antiquities of the New. Only a few copies of his book are to be found now in libraries here and abroad, and this English translation is the first, to the best of my knowledge, to be published. Although Albert C. Koch wrote letters to Germany and kept diaries describing his travels, these writings seem to have been lost. Some of Koch's diaries and letters were the basis of a four-volume work by Moritz Beyer, a German who had traveled in the United States, but whose own notes had been lost in transatlantic shipment. Beyer was a friend of Albert C. Koch and his brother, Louis Koch; because Koch's diaries coincided in time, and in places visited, with his own travels, he used them to refresh his memory. The four volumes were published by Beyer in Leipzig from 1839 to 1841 in collaboration with Louis Koch under the title *Amerikanische Reisen* [American travels].

It is not the purpose of this Preface or of the Introduction which follows to justify the geological, anatomical, or other scientific theories of the writer of this journal, nor to prove his critics wrong or discredit them either as scientists or human beings. After all, this is a travel diary, and the bibliography and footnotes provided later should satisfy, I hope, the more curious and scientific-minded who wish to pursue such subjects in depth.

The writer of this diary had the infuriating habit of mentioning usually only the last names of people he met, thereby making it very difficult for the editor to identify these persons. Moreover, as a German, he heard the names of places and persons just as they were pronounced, and he spelled them according to German rule. The transliterations indulged in by the writer proved at times very disturbing and vexing, but also provided occasional exhilaration when, after hours of staring at a certain word, inspiration suddenly struck and the right expression came to light. As an example, Koch talks about "Handeisen," which in German are handcuffs, or, in literal translation, "hand irons." These, after some head pounding, finally turned out to be "andirons."

Koch's literary style, with its maddening, rambling, continuous half-page sentences of ponderous nineteenth-century German, might best be evaluated in his own words, when he says: "It is a truly enchanting view which I would like to have described by a more skilled pen."

I should like to express here my gratitude to one who is not only an eminent scholar but also an American who has been very kind to a newcomer from the Old World and who has entrusted him with the present undertaking. I am deeply indebted to Professor John Francis McDermott, Research Professor of Humanities at Southern Illinois University, the spiritus rector and general editor of this series, Travels on the Western Waters, for his guidance, patience, and understanding help.

Koch has mentioned a number of people, places, and events which I have tried to identify or clarify. I could not have done this without the help of many people and institutions, and I am deeply grateful for information supplied and assistance rendered by Mr. Milo B. Howard, Jr., Director, Alabama Department of Archives and History; Mrs. Frances B. Macdonald, Manuscript Librarian, Indiana State Library; Miss Caroline Dunn, Librarian, William Henry Smith Memorial Library, Indiana Historical Society; Mr. Samuel W. Thomas, Director of Archives and Records Service, Jefferson County Archives and Records Service, Louisville, Kentucky; Dr. Carl H. Chapman, Professor of Anthropology, University of Missouri; Mr. Charles van Ravenswaay, Director, Henry Francis Du Pont Winterthur Museum; Miss Marion B. Kelly, Erie (Pennsylvania) Public Library; Mrs. Katherine V. W. Palmer, Director, Paleontological Research Institution, Ithaca, New York; Mr. John L. Lochhead, Librarian, The Mariners Museum, New-

xv

port News, Virginia; Mr. James J. Heslin, Director, The New-York Historical Society, New York City; Miss Elizabeth Tindall, Reference Librarian, St. Louis Public Library; Mrs. Nicholas Joost, Elijah P. Lovejoy Library, Southern Illinois University, Edwardsville; Miss Ruth Ferris, Miss Elizabeth Golterman, Mr. Frank Magre, Crystal City, Missouri; and Mr. George Brooks, Director, and the entire staff of the Missouri Historical Society, St. Louis.

In the German Democratic Republic my thanks go to Rat des Kreises Bitterfeld (Bezirk Halle) and to Dr. K. Fischer, Museum fuer Naturkunde, Humboldt Universitaet, Berlin; in Austria, to Dr. Heinz Kollman, Naturhistorisches Museum, Vienna; in the Federal Republic of Germany, to Miss Edith Chorherr, Librarian, Bavarian National Museum, Munich; and in England, to Mr. M. J. Rowlands, Librarian, General Library of the British Museum (Natural History), London. Particular thanks for most patient help and vital information are due The Honorable R. Gerald Trampe, Associate Judge, Circuit Court of Illinois, First Judicial Circuit, and Mr. William Hoffmann, both of Golconda, Illinois. A special measure of gratitude goes to my wife, who typed the manuscript and lent invaluable editorial assistance, and without whose patient help I could not have undertaken and completed this project. Being translator and editor in the same person deprives me of any excuse to shift the blame for any mistakes or omissions.

St. Louis Missouri
May 1972

Ernst A. Stadler

Introduction
By Ernst A. Stadler

When I arrived home at noon I was drenched quite thoroughly by the rain and sweat, and had to change my clothes completely. If somebody had painted me in the getup in which I came home, it would have made a queer picture. On my back I had my heavy sack with the oysters; not only my jacket, but also my vest pockets were filled with fossils; under my arm I carried the umbrella (which I could not open because I had both hands full), in one hand the stone hammer and chisel, in the other a handkerchief full of petrified starfish; and I was covered from head to foot with sand and dirt.[1]

THE RAMBLING rock-pile who wrote the above lines might have been seen 125 years ago hanging from cliffs at Gay Head on Martha's Vineyard, probing islands in the falls of the Ohio River, marching along the banks of the Mississippi looking for petrifactions while keeping up with a slow-moving steamboat which carried his baggage, or digging under a hot sun in an Alabama hollow for the remains of a sea serpent, with only a straw mat to sleep on and salt pork and bread for food.

This strange creature was a German geologist and fossil collector by the name of Albert C. Koch. He was born May 10, 1804, in Roitzsch, a village of 244 houses, 1300 inhabitants, and six taverns, in the Duchy of Saxony. The village comprised a royal domain, four manorial estates, and one parochial estate. The villagers owned twenty-five hides [2] of land, on which they grew all kinds of grain, especially fine wheat, beets, and some tobacco. There were many wool carders and peddlers in the village, and the poorer villagers made a living knitting socks, an occupation for which even children from the age of five years on were used. In these surroundings, steeped in almost medieval conditions, young Koch grew up. His father, Johann Eusebius Sigismund Koch, was magistrate and administrator of the royal domain. Koch, senior, had established in his house a cabinet of natural history, in which was included a collection of shells but which excelled particularly in stuffed birds.[3] Albert C. Koch's interest in natural science, as well as his consuming urge to travel, seems to have been inherited from his father. Of Koch's early childhood and youth nothing is known to us so far. It can be assumed, judging from his father's position, that he received at least a secondary education, but there is no record of attendance at a university. Although Albert C. Koch, who after 1845 carried a "Doctor" in front of his name, wrote extensively about his travels and discoveries, he kept almost entirely quiet about his personal life.

His pamphlets describing his discoveries encompassed usually up to thirty pages and, except for a few in German, were written in English. His longer works are all in German, his diary running to 162 pages, and *Die Riesenthiere der Urwelt* [The giant animals of the primeval world], to 99 pages. This latter work is a recounting of his discovery of the *Missourium* and a general discussion of the mastodon family. His third major work, *Die Sechs Schoepfungstage oder die Mosaische Schoepfungsgeschichte* [The six days of creation or the Mosaic history of creation] is a 51-page treatise in which it was Koch's intention to prove that the day-by-day account of the creation of the world as told in the first book of Moses is in exact conformity with geology. This book, printed in Vienna in 1852, was, not surprisingly, dedicated to his eminence the Prince-Bishop, Cardinal Melchior von Diepenbrock of Breslau.

Information about Koch's early years in America, until he established himself in St. Louis, can so far only be gleaned from his obituary and his diary. From his obituary we learn that he crossed the Atlantic Ocean when he was twenty-two years old. Where he landed after his arrival in 1826, where he stayed, or how he earned his living in those early years we do not know. Koch's obituary also mentions that "very interesting details concerning Dr. Koch's exhumations in Missouri, will shortly be published in the scientific journals, from a posthumous autobiography." [4] No trace of any such biography has been found.

The only clue to Koch's whereabouts in those early days is found in the *Anzeiger des Westens,* St. Louis's first German newspaper, which mentions on June 29, 1839, that Koch had been a citizen of St. Louis for 12 years, which would mean that he came almost directly to St. Louis from Germany.[5] If he did, he probably did not stay very long because in his diary is found an entry that he was in Erie, Pennsylvania, in 1830.[6] The obituary of his son, James Albert Koch, mentions that Albert C. Koch married in Erie a Miss Reid of Philadelphia, and that afterward he had moved to St. Clair, Michigan, where two daughters, Rosalia and Maria, were born.[7] St. Clair records have unfortunately not divulged any information about Koch's stay there. Koch's presence in St. Louis can only be satisfactorily established as of 1836; in that year his name appeared in newspaper advertisements as the proprietor of the St. Louis Museum.[8] In 1837 his son James was born, and in 1838 his residence was sufficiently established to allow for listing in the St. Louis directory as museum proprietor.[9]

Albert Koch's reign as the owner of the St. Louis Museum was marked by imaginative and lively innovations. The visitors to the museum at Market and Second streets were treated to striking wax likenesses of President Andrew Jackson and General Don Antonio Lopez de Santa Anna, among others; there were rare stuffed birds of distant regions, and Koch had not neglected to put their eggs right at their sides. There were also "cosmoramic views" of battles and ancient cities, and an exhibition of the infernal regions.[10] The *Missouri Saturday News* of January 20, 1838, remarked of the exhibition: "In fact, the whole illustration of natural history, consisting of Beasts, birds and creeping things, is in very good keeping, and evident of great talent and industry."

Again, on May 26, 1838, the *Saturday News,* obviously quite taken by Koch's museum, stated: "Mr. Koch in his untiring efforts to please his visitors, is really a deeper practical student of natural history than he has been hitherto considered. The recent additions which he has made to his collection of living animals, are exceedingly attractive and worthwhile." The writer then speculates that the live grizzly bear might fight one of the five alligators there: "The largest alligator measures fifteen feet in length, and he is a thrifty specimen of the lizard family. There is an irreconcilable misunderstanding subsisting between this personage and the grizzly bear; when the meeting takes place, we understand seasonable notice will be given. To whichever side victory may incline in the coming combat, it is presumed very little of either belligerant [*sic*] will be left when the battle ends." There is no record that such a battle ever took place. The alligators, however, did fight fiercely among themselves. One of the creatures, perhaps out of fright or panic, either threw itself or fell from a third-story window to the sidewalk below. On June 2 the *Saturday News* could then report under the heading "Suicide" that "some personal conflict had occurred at the Museum in this city a few nights since," and, as a result, "that a dead Alligator lay in a public street of the city."

Another one of Koch's curiosities, a painting of the famous Seminole chief Osceola, played a part in a controversy between the editor of the *Saturday News* and George Catlin, celebrated painter of Indians. On February 10, 1838, the *Saturday News* printed under the headline "High Coloring—More Vermillion," the following letter:

> Fort Moultrie, Charleston
> 17 Jan., 9 o'clock
> Dear Sir. I arrived here this morning in three days, fine weather, and well. I have just had an interview with Oseola [*sic*] and other chiefs—had a talk with them, and begin in an hour from this to paint.
> I shall paint Oseola, Coachajo, Micanopy, Cloud, King Philip and several others, and hasten back with all speed to show the citizens of New York how these brave fellows look.
> You will think by this time that I am catering for the world at great expense to myself, and it is even so, "but things to be done, must be done."
> Oseola is a fine and gentlemanly looking man, with a pleasant smile that would become the face of the most refined or delicate female—yet I can imagine, that when roused and kindled into action, it will glow with a hero's fire and a lion's rage. His portrait has never yet been painted. In haste yours. Geo. Catlin

Catlin's letter seems to have aroused the ire of the *Saturday News* editor, who took great pleasure in adding his own caustic evaluation of Osceola and Catlin's sentiments. The editorial reply reads:

> The portrait of Oseola has been painted; and much coloring already employed about his person. At the commencement of the Florida war, he was made a perfect beauty-spot with grease and lamp-black, with which his "refined and delicate" face was then besmeared.
> In St. Louis we have been for a long period accustomed to look at the portrait of Oseola. Mr. Koch the proprietor of the St. Louis Museum, who

is not anticipated by any artist in the Indian line of business, has placed a full length portrait of this "gentlemanly lion" in transparency in front of his museum. When the rats were apprised of the location of the hero's picture there, they left the building in wild affright, and never returned for their baggage.

The "refined and delicate face" of this hero, when lit up at night, has more than once made deep and lasting impression on the hearts of the mulatto girls who are accustomed to pass the museum.

Large contributions should be levied to indemnify the artist for his patriotic efforts to set out in bold relief such "delicate and refined" beauties as Oseola is; and humanity should not be left to deplore the pecuniary suffering of a man who has already done so much to encourage painted red men to heroism in taking our scalps. Could not the artist contrive to elicit for Oseola sufficient interest to secure him a triumphal entry into the commercial emporium? Under the influence of Heathen mythology, Oseola and the artist might have been both deified—one with the pencil, and the other by the tomahawk.

On August 11, 1838, a writer for the *Saturday News* with the improbable name of Felix Lurid, reporting from rural Missouri, appended the following note to his column: "The young Proc which I send by this day's mail raft, I fear will not survive the voyage, but the ingenious Mr. Koch can preserve the skin, and thus exhibit the form of this singular animal." On the same page of the newspaper appeared a letter from Koch acknowledging the receipt of the animal. What was apparently intended as a practical joke played on Koch was turned by him into an eye-catching addition to the museum, for on September 3 there appeared an advertisement in the *Daily Commercial Bulletin* which offered the new animal under the fetching caption "Did You Ever See a Prock?" Because the ad also gives us an interesting glimpse of other attractions in the museum, it follows here in full.

DID YOU EVER SEE A PROCK?

This beautiful animal which has long been considered fabulous, is now to be seen at the *St. Louis Museum*.

Not in a painting, but stuffed and preserved as natural as life, situated on a rock surrounded with flowers—The Animal is the size of a mule and has much the resemblance of a Zebra, except that the legs are much shorter and thicker and the tail more bushy at its extremity. The head has much the resemblance of a Rhinoceros, and has a horn on the nose of an ivory substance. The public are requested to call and examine the animal for themselves, as the limits of an advertisement are insufficient for a full description. Call and behold an animal that has never before been exhibited in the old or new world.

ALLIGATORS

By this opportunity the proprietor of the St. Louis Museum has the pleasure to announce to his friends, the ladies and gentlemen of St. Louis and vicinity, the reappearance of the two largest of the Alligators which created so much excitement in the fore part of the summer. These are the alligators which had a tremendous battle, in which the smallest, having been overpowered by his antagonist, broke through the window, leaped over the iron balcony in front of the museum and broke his neck; the other died a few days after, in consequence of the wounds received in the fight.

Both are represented in the attitude of fighting—blood flowing from their wounds, their jaws locked in a deathly embrace, and the whole representing the ferocious nature of the animals.

WAX FIGURES

Among the wax figures are likewise several alterations and improvements, but we have only room to mention the reappearance of the Seminol Chief OSCEOLA and his wife, dressed in full Indian costume.

INFERNAL REGIONS

The Evil Spirit, who is never at rest, has also been busy, and has lately received some new recruits which has not, of course, made the place more handsome, but has contributed to its horrific appearance. It will positively be exhibited every evening at 9 o'clock.

Museum, corner of Church and Market streets. Admittance 50 cents
Children half price

The director of the museum also doubled as a theatrical impresario of note who was offering his public a wide variety of entertainment. So at times his museum was turned into a theater where his audience was regaled by ventriloquists, magicians, bird imitators, singers, and actors. Comic songs and plays like *The Swiss Cottage, or why don't she marry* were among the offerings.[11]

While the museum provided Koch with the necessary means for making a living, it also gave him the time to travel in Missouri and to devote his energy to his first love, the collecting of fossil bones and petrifactions. And here he did truly astonishing work considering that he was not a trained or professional geologist or paleontologist, as some of the scientists of those and later times were eager to point out. What he might have lacked in pure scientific knowledge he more than made up by perseverance and a seemingly inexhaustible energy in the quest for earth's hidden treasures. While in later years some of his more learned colleagues disputed some of the conclusions he had drawn from his unearthed animals, there was never anything in their remarks but admiration and praise for his untiring work. The *St. Louis Daily Evening Gazette* of April 30, 1840, gives us a glimpse of the man in a short report:

KOCH'S RESEARCHES

All who have ever visited the St. Louis Museum, know that it is kept by a plain man, without a bit of pretence, affectation or quackery in his composition. His ardor as a virtuoso is equally remarkable as the singleness of his mind and simplicity of his manners. Let Mr. Koch hear that a strange fossil has been exhumed within three or four hundred miles of his residence, and he is sure to be there. No man has been so indefatigable—no man has expended half as much money, time and labor, in exploring and bringing to the light of day the remains of antiquity which have been buried for uncounted centuries in the bosom of our forests and prairies.

A few weeks ago we met him on the steamer *Leavenworth*. He was then on his way to a spot near the Missouri River, about 60 miles distant where he learned that the fragment of a strange bone had been discovered. He was sanguine in the hope that he should find more remains; and thus complete the skeleton of an entire Mastodon. More recently we accidentally met him again on another boat; and now found that he was bound on an exploring expedition to the "Osage Country"—beyond Benton County; the spot was not exactly designated but perhaps a matter of two or three hundred miles from the city. He had not been successful in the trip, we first spoke of, a few bones only rewarding his search. Now, however, he had reason to believe, upon very strong assurances, that he should be able to unearth the "entire animal."—At any rate, he was impatient to get to its locality, before ignorant and unscientific hands had an opportunity of disturbing the sacred relics.

All this time Mr. Koch has been laboring under an attack of fever and ague, contracted in his several exploring expeditions. But he is not a man to be daunted by such trifles. We hope his present fatiguing journey will be crowned with the complete success, which he anticipates. He certainly deserves it.

The hopes expressed by the newspaper for complete success of the indefatigable fossil hunter did indeed come true. While he had found and disinterred in 1838 and 1839 the remains of various animals, it was not until May 1840 that he actually discovered the almost complete skeleton of a mastodon which would forever be linked with his name, and which he named in honor of the state whose earth had yielded the treasure.

On March 24, 1840, having just returned from a trip up the Missouri River, lying in bed racked with fever, Koch received a message that a farmer in Benton County had found some fossil bones while building a mill. Finding himself in a dilemma, Koch wrote: "Now reason fought with the love I felt for my explorations. The former pointed out that my health was already shattered by the many previous fevers, serious and slight, which I had contracted particularly by inhaling unhealthy swamp air. The latter, on the other hand, painted for me a fantasy which afforded me the prospect of a yield such as I had not yet had, and which naturally could not be mine if I did not overcome my physical weakness and go without delay on the long and arduous journey."[12]

Twenty-four hours later, on March 25, Koch was on a steamboat going up the Missouri, hurrying toward his destination. It took him six days to reach the spot where he would cross the Osage River, swollen, rapid, and deep from the spring rains, in a small boat with his guide, their horses kept close by the boat. After a 24-mile ride through virgin forest they reached the Pomme de Terre River, where, no boat being available, they were forced to ford the river on their horses. In a romantic valley some 30 miles from the spot where the Pomme de Terre enters the Osage Koch took his quarters at the log cabin of the owner of the land and started his diggings, which would keep him there for four months. His first group of workmen left after only two days, and the second also found the work too hard and soon followed the first; finally, a third crew was assembled which stayed and finished the job.

Koch had originally planned to ship his treasures down the Pomme de Terre into the Osage, and on the Osage to the Missouri, where they could be loaded onto a steamboat for St. Louis. For this purpose he had felled four large black walnut trees and had them hollowed out, Indian fashion, and made into canoes. These he then joined together with beams. But when the canoes were ready, the Pomme de Terre had fallen so low that the dugouts were useless. Although he had dreaded the thought of transporting his treasures overland, and considered it only as a last resort, now there was no other way. He hired three large wagons, each drawn by four oxen. The sideposts of the wagons he had replaced with hewn slender young and elastic oak trees which acted as springs and from which he suspended his wooden boxes containing the bones of the skeleton. This caravan made its way slowly to Boonville, where the precious load was put on a steamboat for St. Louis.[13] By August of the same year Koch had assembled the contents of his boxes, and from them emerged an astounding creature which would in the years to come be known variously as the *Missourium*, the *Missouri Leviathan*, or the *Mastodon giganteus*, until it found its final place as the *Mastodon americanus* in the British Museum.[14]

Because Koch had no other resources to finance his digging expeditions than selling some of the fossil bones and exhibiting his prehistoric creatures, he would extol his exhibitions with considerable finesse in newspaper advertisements. So the St. Louis public was exhorted in a long advertisement in the *Daily Commercial Bulletin* of August 25, 1840: "Citizens of Missouri, come and see the gigantic race that once inhabited the space you now occupy, drank of the same waters which now quench your thirst, ate the fruits of the same soil that now yields so abundantly to your labor. To you stranger, I say, come and see the wonderful productions of unknown ages which are no where to be found but in the humble abode of the St. Louis Museum."

The skeleton the museum visitors saw was thirty-two feet long and fifteen feet high. The reason for its so very remarkable size was that Koch in his enthusiasm had put together twenty-three dorsal vertebrae instead of the correct nineteen, and instead of four lumbar vertebrae, he had ten. Moreover, some ribs were superfluous, and the spaces between the vertebrae were filled with wooden blocks thicker than the original cartilage which had held the skeleton together.[15] That errors should occur is not surprising, considering that some of the most prominent paleontologists and geologists had to—and still have to—reconsider and correct some of their evaluations in the light of new discoveries.

In January of 1841 Koch sold his museum and took his prize, the *Missourium*, on an extended tour of various cities prior to shipping it to Europe. The *Missouri Republican* of February 4 reported that Koch

had left with his *Missourium* for New Orleans, and expressed the hope
that some friends of science might take measures to keep the huge
monster in the United States. In October 1841 the *Missourium* was
exhibited at the Masonic Hall in Philadelphia. In the magazine *The
Farmers' Cabinet*, published there, appeared a three-page description
of the skeleton written by Koch and accompanied by a drawing of the
beast.[16] Dr. Paul B. Goddard [17] of the Academy of Natural Science in
Philadelphia had reported on the deficiencies of the skeleton, and in
the pages of the magazine there developed a lively exchange of opinions
between opponents and supporters of Koch. At the October 15, 1841,
meeting of the American Philosophical Society of Philadelphia, to
which Koch had sold a collection of fossil bones in 1840, Dr. Richard
Harlan [18] read a memoir commenting on Koch's exhibition: "There is
now exhibiting at the Masonic Hall, in Philadelphia, one of the most
extensive and remarkable collections of fossil bones of extinct species
of mammals which have hitherto been brought to light in this country,
a gratification for which our scientific community will acknowledge
themselves indebted to the perseverance of the enterprising proprietor,
Mr. Albert Koch of St. Louis , Missouri." In the 12-page study, Harlan
examined the skeleton and called attention to some of the errors, of
which he said: "The proprietor not possessing the advantage of anatomical knowledge has committed some errors in the articulation of the
bones, which, no doubt, his ulterior researches will enable him to rectify." [19]

Koch did not seem to have had the time or inclination to rectify some
of the errors before he left for Europe. When he showed his *Missourium*
late in 1841 at the Egyptian Hall in Piccadilly, London, creating great
curiosity, the noted anatomist Richard Owen, at a meeting of the
Geological Society of London on February 23, 1842, pointed out certain
mistakes in the makeup of the skeleton and mentioned the willingness
of Koch to correct them. But he differed sharply with Koch on the
classification of the *Missourium*.[20] Some time later Koch answered
Professor Owen at a meeting of the same society, and later in that year
Dr. Robert E. Grant of the University of London came to Koch's
support.[21] All through 1842 the *Missourium* was to be seen in London,
and, according to a pamphlet prepared by Koch for his exhibition, the
skeleton was shown in Ireland in 1843.[22] In that same year Koch also

attended a meeting of the natural history association, Isis, in Dresden, and in November the *Missourium* was purchased by the British Museum for the sum of £1300.[23] The Royal Museum in Berlin also acquired some of Koch's fossils and some Indian implements found with the *Missourium*.[24]

Professor Owen was later entrusted with the reconstruction of the *Missourium*. The famous paleontologist Gideon Algernon Mantell [25] wrote in 1848 in a letter to the *Illustrated London News*, in which he criticized Koch's sea serpent, "Mr. Koch is the person who, a few years ago, had a fine collection of fossil bones of elephants and mastodons, out of which he made up an enormous skeleton, and exhibited it in the Egyptian Hall, Piccadilly, under the name of *Missourium*. This collection was purchased by the trustees of the British Museum, and from it were selected the bones which now constitute the matchless skeleton of a Mastodon in our National Gallery of Organic Remains." [26]

While Koch must have enjoyed his stay in London and the attention which his discovery received there, one wonders what effect all this must have had on his family. The children were put in school in Dresden while Koch and his wife traveled on the Continent and in England, where, during their stay their youngest child, Georgena, was born in London.[27] Soon Koch was preparing for another trip to America, leaving his wife and children behind in Germany. Apparently freed, at least for a while, from money worries by the sale of the *Missourium* and other fossil remains, Koch embarked in 1844 on his most ambitious project, a two-year trip to the United States which would take him from Gay Head on Martha's Vineyard to New Orleans and finally to Alabama to search for his sea serpent. He kept a full day-by-day account of his travels, of which excerpts were published in Germany in 1847. This volume, here translated, not only shows Koch as a dogged hunter in pursuit of his beloved fossils but also gives us some fascinating descriptions of travel in the mid-nineteenth century as well as observant comments on customs and places in the United States.[28]

Koch's impressions of his two-year wanderings range from a dry layer-by-layer cataloguing of rock formations to lyric descriptions of a New England town, to an enumeration of food items (especially for the benefit of his German readers) of an American evening meal and a breakfast staggering in its variety, and to comments on the lack of imagination of German building officials when he sees how Americans move houses from one location to another. He theorizes that German officials would hardly have hit upon such an idea. His often emerging and probably not always endearing quality of relentless obstinacy in attaining his goal is, among other examples, well illustrated in his account of an encounter with a snake which he attacked with Teutonic furor by hurling his water bucket and knife at it. Characteristically, he wrote: "I could not leave without having shown the snake that I had proved already to many of its peers that it would be best to run off when confronted by a hereditary enemy superior to them." Neither the snake nor the indefatigable fossil hunter pressed his point, however, and each went his own way.

The presidential election campaign of 1844, which he witnessed in Kentucky and Indiana, was bewildering to Koch—as it would have been to many other foreigners—but he nevertheless gives us some vivid descriptions of the strange goings-on.

While Koch was on his way to Alabama in December 1844, the steamboat on which he was a passenger was caught in an ice jam near the Grand Tower of the Mississippi and had to be abandoned by the passengers in the early morning. Koch could not tear himself away from his boxes of petrifactions and stayed on board until noon, when he finally left the boat. Unlike his fellow passengers, who were happy to have escaped with their lives, Koch could not think of giving up his boxes, and he was able to persuade a farmer to take him on a night ride in an oxcart back to the spot where the boat still lay in the ice. With the driver and two helpers breaking through the ice to the boat, Koch managed to get the boxes safely on land. He then made his way inland to Golconda, Illinois, where he discovered lead deposits which he thought showed promise of great fortune. He bought land there and envisioned himself as owner of one of the richest lead mines in the country. While his dream of the rich lead mines never materialized, he did acquire enough land there to be able to put it to good use in later years.

In January of 1845 Koch arrived in Alabama where he stayed for three months visiting Macon, Clarksville, Coffeeville, St. Stephens, and Old Washington Courthouse. On foot and on horseback he covered the land and climbed over the banks of the Tombigbee River, the Alabama River, and Tattilaba Creek looking for fossils and for the great sea serpent he had been dreaming about. He found various ribs and vertebrae, but not until April did he discover the almost complete *Zeuglodon* skeleton. In his diary he describes the travails of unearthing and transporting the monster by steamboat to New York.

By August of 1845 Koch had finished assembling the sea serpent and was exhibiting it in the Apollo Rooms on Broadway. It was a wondrous creature that emerged from the boxes of the inveterate fossil hunter, and its length of 114 feet created quite an impression, especially on the gentlemen of the New York press. In a twenty-four-page pamphlet prepared for the exhibition, the exhibitor appeared for the first time as *Doctor* Albert C. Koch. Of the sea serpent he wrote: "This relic is without exception the largest of all fossil skeletons, found in either the old or new world. Its length being upwards of one hundred and fourteen feet, without estimating any space for the cartilage between the bones, and must, when alive, have measured over one hundred and forty feet, and its circumference probably exceeded thirty feet, reminding us most strikingly, of the various statements made by persons, in regard to having seen large serpents in different parts of the ocean, which were known by the name of Sea Serpents." At the end of the pamphlet Dr. Koch, always the astute public relations man, had reprinted long comments from the New York papers on his exhibition.

The journalists were fulsome in their praise of the magnificent beast and indulged in hyperbolic descriptions and speculations as to its nature and origin. The *New York Dissector* wrote: "The serpent of the Deucalian deluge, slain by Apollo Pythius, is beheld, with scarcely the aid of the dullest fancy, in the Apollo Saloon in Broadway." The *New York Evangelist* said: "There is great grandeur in the manner in which he carries his jaws. And his eyes doubtless fired up in the green sea-depths like great lamps in a cavern, or like the lights by night in front of railroad locomotives. And if we might figure a wonder of earth, which Milton, if he had ever seen it, would have made use of in some way or other, to which this creature may be likened in the sea, it would be an immense train of cars, with steam engine and all, shot into the bowels of the deep, and flying through the water at the rate of sixty miles an hour." Later in the text the same writer asked: "Who knows but he had seen the Ark? Who knows but Noah had seen *him* from the window?

Who knows but he may have visited Ararat? Who knows how many dead and wicked giants of old he had swallowed and fed upon? Perhaps, when we now touch his ribs, we are touching the residium of some of Cain's descendants, that perished in the deluge." And, to give the reader and viewer some more food for thought, the paper at the finish printed the forty-first chapter of Job, admonishing the readers to take the chapter with them when viewing this wonderful skeleton; if they could not see or imagine in Koch's serpent the monster leviathan described in it, it must be because they could see nothing but dry bones in anything. The paper also asserted that the *Hydrarchos* of Dr. Koch should be retained in this country at all costs. A writer for the *New York Morning News* ventured a guess about the eating habits of the beast with the estimate: "Three buffaloes at a meal must have been about a fair allowance for this capacious feeder, who plumply realizes in bulk the portrait which history has drawn of the Great African Snake which stopped the march of the army of Regulus, and yielded at last only to the ponderous rock hurled at him from the Catapult." [29]

Scientists who saw and examined the sea serpent while it was on exhibit in New York and Boston criticized its exaggerated size and pointed out that the skeleton was composed of parts of more than one individual, belonging to different genera of animals, and also established the fact that it was not a reptile but a mammal. Skeletons and remains of skeletons similar to that found by Koch had been found previously in Alabama, and scientists had described them variously as *Basilosaurus* and *Zeuglodon*, while Koch named his monster *Hydrarchos*.[30] Because he was not satisfied with the scientific evaluations of his find in this country, he felt it advisable to take it to Europe and lay it before the celebrated anatomists and naturalists, Professor Carl Gustav Carus of Dresden [31] and Professor Johannes Mueller of Berlin.[32]

Koch embarked on June 6, 1846, for Hamburg, where he arrived one month later. In Hamburg he entrusted his *Hydrarchos,* in the care of a shipping agent, to be sent on to Dresden, while he eagerly hurried ahead to see his family after so long an absence.[33] The skeleton was shown in 1847 at the Leipzig Fair and later in Berlin. King Friedrich Wilhelm IV of Prussia, called "the friend of natural scientists," had Koch's collection bought for the Royal Anatomical Museum in Berlin. The sales contract, dated May 19, 1847, gave Koch a yearly pension for his lifetime.[34] After the collection had been placed in the museum, studies by Professor Mueller confirmed some of the criticisms made by the American scientists.[35]

In January of 1848 Koch was back in his old haunts in Alabama digging for more bones, and on the seventh of February he found another *Zeuglodon* skeleton. After working on it for several months he was able to ship the skeleton, still partly encased in its native lime rock, to Dresden. There, eight months later, the skeleton, measuring ninety-six feet, was freed from its rock encasement and on May 6, 1849, Koch exhibited this second *Zeuglodon* in a large and beautiful saloon furnished by the Royal Academy of Dresden, where it was viewed by the whole royal family of Saxony. This exhibition was followed by one in the city of Breslau, from whence he took his prize skeleton on tour to Vienna, where he stayed a year before going on to Prague. A visit to Munich did not materialize because no rooms large enough for his collection could be found, and because some urgent business required his presence in the United States, so he said, "Thus I concluded to take the *Zeuglodon* back to its native country, after it had established its just fame in Europe." [36]

In 1853 Koch appeared with his *Zeuglodon* in New Orleans, where William T. Leonard, manager of the museum on St. Charles street, announced that Mr. Dan Rice had engaged this immense antediluvian monster for a showing in his Great Southern Museum.[37] From New Orleans Koch went to St. Louis, and the *Zeuglodon* was purchased by Edward Wyman, owner of the St. Louis Museum. In 1863 the skeleton was acquired by Wood's Museum in Chicago.[38]

In the mid-fifties the peripatetic paleontologist ceased his wanderings and settled with his family, consisting of his wife, three daughters, and a son, in St. Louis, where his son James attended Washington University as a medical student. The St. Louis directory for 1854–55 lists Albert C. Koch as a professor of philosophy, residing at 68 South Sixth Street. In April of 1856 Koch was elected an associate member of the Academy of Science of St. Louis. In May of the same year, having persuaded the Academy to foot the bill for a trip to Mississippi, he was busy procuring for the Academy the remains of a *Zeuglodon* and other fossil remains.[39] In 1857 he was elected one of the four curators of the Academy. It was in 1857 also that he felt compelled to present again to the public and the scientific world, in the *Transactions of the Academy of Science*, his theory of the contemporaneity of man with the mastodon in North America.[40] Although Koch had as early as 1839 mentioned his discovery of human artifacts in connection with extinct fossils, the discovery was overshadowed by his mastodon finds, and very few scientists until 1971 ever acknowledged it.

From 1857 to 1861 Koch was also coeditor with his son-in-law, Robert Widmar, of the *Mississippi Handels-Zeitung*.[41] M. Hopewell's *Report of the Fourth Annual Fair of the St. Louis Agricultural and Mechanical Association*, September 1859, mentions that "Dr. Albert Koch of St. Louis, received a diploma for his exhibition of Platina ore." [42]

Albert Koch had an inquiring and imaginative mind, but he was also a practical man who realized that without money he could not search for his fossils. It was this very practical feature which caused him to be accused of being more of a fossil merchant and showman than a true scientist. It is not surprising that some of his scientific colleagues, with institutions and professions as backing, should frown upon his mercantile exploits. In the mid-1860s Koch entered with George Babcock, a St. Louis attorney, into geological explorations in Tennessee for iron, lead, coal, and petroleum, and the Knoxville Oil and Mining Company was formed on the recommendations of the two explorers.[43] During these same years Koch had also engaged in real estate transactions in Golconda, Illinois, where his brother Louis played an important part in the development of the town. A large section on the south side of the town is still known as "Koch's addition to the town of Golconda." [44] It was here, at the country estate of his brother Louis, that Albert C. Koch died of a "lingering torpor of the liver" on December 28, 1867.

After his death Koch became almost a forgotten man, although some scientists did from time to time try to disprove and disparage his theories on his discoveries calling them Munchausenian exaggerations and including his name in books dealing with scientific hoaxes and frauds. It is interesting to note, however, that one of the foremost American geologists, James D. Dana, who was one of Koch's severest critics, used Koch's *Missourium*, as reconstructed by Owen, as an illustration of the *Mastodon giganteus* in his *Manual of Geology*.[45] At this point come to mind the words of Charles Darwin, who said in his *Descent of Man*: "False facts are highly injurious to the progress of science, for they often endure long; but false views, if supported by some evidence, do little harm, for everyone takes a salutary pleasure in proving their falseness; and when this is done, one path towards error is closed and the road to truth is often at the same time opened." [46]

John W. Foster, American geologist and paleontologist and president of the American Association for the Advancement of Science, who had known Koch personally, wrote in his *Pre-Historic Races of the United States of America* in 1873: "He [Koch] was an indefatigable collector, and few men in this country, by individual effort alone, have been more successful in bringing to light so many of the skeletons of the huge animals that roamed over the land or swarmed the seas of past ages. His knowledge in many branches of Natural History was considerable, but not of that exact character to bring out important generalizations. No one who knew him will question but that he was a competent observer, and to deny the accuracy of his statement is to accuse him of having attempted to perpetrate a scientific fraud. It may be said, however, that the scientific opinion of this country regarded his statements in about the same light as the French geologists did those of M. Boucher de Perthes,[47] when he brought out his *Antiquités Celtiques,* that is with absolute distrust: but the one lived to see the truth of his observations acknowledged and their value appreciated; the other died with a cloud hanging over his reputation." [48]

Koch's theory of the existence of man in association with extinct fossil animals in North America, mentioned by Foster, has gained more credence in the intervening years with the great increase in new knowledge in the field of geology. In 1944 the eminent anthropologist M. F. Ashley Montagu, in collaboration with C. Bernard Peterson, wrote in a study of Koch's discoveries and theories: "In this, the earliest account of the association of human artifacts with the remains of fossil mammals in North America, we have, oddly enough, one of the clearest and best evidences of the antiquity of man in North America that has ever been published. Yet Koch's statements have nearly always been dismissed as unworthy of belief. He has been ridiculed and completely shut out of court. None of his critics has ever taken the slightest trouble to check any of his findings, although Koch clearly stated where his materials were to be found, even after he had sold them. No one has ever examined his statements by checking them against the findings and discoveries of later investigators, discoveries which in every detail assist to increase the probability that Koch discovered what he claimed to have discovered." [49] Professor Carl H. Chapman of the University of Missouri wrote in 1964: "In fact, his data are more acceptable today than they were then, for there is not so much prejudice against them." [50]

New evidence has been recently discovered supporting Albert C. Koch's claim that Indian artifacts were associated with mastodon bones at his excavation sites. In the summer of 1971 Dr. R. Bruce McMillan, Associate Curator of Anthropology at the Illinois State Museum, and Dr. Everett H. Lindsay of the University of Arizona, among a group of scientists, began, with the aid of a National Science Foundation research grant, a reinvestigation of the site in Benton County, Missouri, where Koch made one of his disputed discoveries. Dr. McMillan has very graciously informed me of the results of the investigation, and has given me permission to make use of them here. The scientists who worked at the Koch site found landmarks mentioned by him, and the excavation of Koch's original site yielded considerable scrap mastodon bones in Koch's spoil dirt. Dr. McMillan states, "It was obvious that Koch had essentially completed excavation of the bone bed." The scientists also found a platform of large crisscross rails, many of them walnut, believed to be from Koch's time, which his work gang had used to stand on while working in the waterlogged spring. Koch mentioned that he had had his men cut down walnut trees. Dr. McMillan and his colleagues found that in the feeder of the spring in which Koch dug, Indian artifacts occurred with mastodon bone. The bones found with these archaeological materials are estimated by Dr. McMillan to be 25,000 years old, suggesting that they are several thousand years older than the artifacts. More exacting studies, such as radiocarbon dating, are still being conducted. While Koch's mastodon finds provided some of the first and finest type specimens for the paleontologists, there was always a shadow of doubt hanging over his reputation. Now, more than 130 years after his discoveries, the shadow has lifted and it has been made clear that although Koch perhaps misinterpreted the connection between the mastodon bones and the Indian artifacts found nearby, he did nevertheless find what he claimed he had found. I am sure that this would please the old fossil hunter very much. That I should have had a small part in bringing to light the accomplishments of the immigrant

Albert C. Koch, who made the American earth his own, pleases *me* very much.

Of Koch's discoveries, only the mastodon skeleton survived. It can still be seen in the British Museum, and a few remaining mastodon bones and stone artifacts are still on exhibit in the Berlin Museum. His sea serpents suffered curiously similar fates; the one which he brought back to its native land became a victim of the great fire of Chicago in 1871,[51] and the other, which stayed in Berlin, perished there in the holocaust of 1945,[52] just one hundred years after its discovery in Alabama.

Perhaps now Albert C. Koch and his antediluvian monsters, forgotten and unappreciated for so long, will receive some of their due recognition, but whether or not due tribute is ever paid to this visionary does not really matter any more, for the ultimate tribute has been paid to him by the same earth whose mysteries he had tried so hard to decipher. When Albert C. Koch's body was moved from its original place of interment, a small burying ground near a vineyard, to the hilltop cemetery overlooking the Ohio River at Golconda, it was found that the mortal remains of the indefatigable fossil hunter had petrified.[53] To how many men is granted the privilege of being so consistent, even after death?

Journey Through a Part of the United States

General view of the mastodon beds at Kimmswick, Jefferson County, Missouri.
This is the general site of Dr. Albert C. Koch's 1838–39 excavations of mastodon
remains, some of which he showed in his museum in St. Louis. Efforts are under
way to make this mastodon bed/site a park. This and the following three
illustrations are made from photographs taken around the turn of the century.
Photograph by George Stark, St. Louis.—Courtesy Missouri Historical Society

"Koch's Hole" at Kimmswick, Missouri, where Dr. Koch found mastodon remains in 1838–39. Photograph by George Stark, St. Louis.—Courtesy Missouri Historical Society

Two mastodon femurs and a humerus found at Kimmswick, Missouri. C. W. Beehler, lessee of the tract, holds jawbone. Photograph by George Stark, St. Louis.—Courtesy Missouri Historical Society

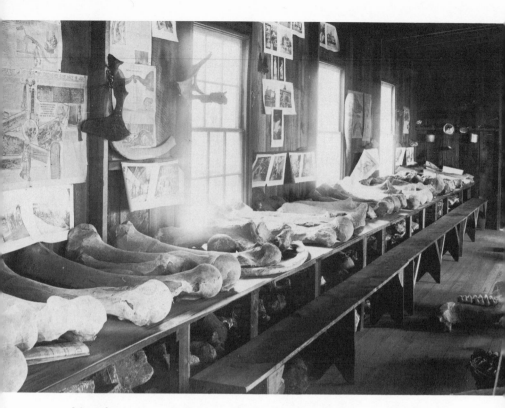

Mastodon remains in the museum of C. W. Beehler, near excavation sites at Kimmswick, Missouri. Photograph by George Stark, St. Louis.—Courtesy Missouri Historical Society

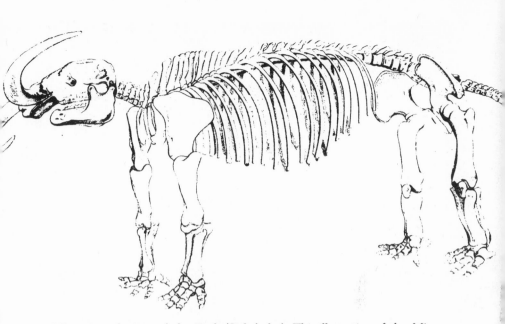

Missourium theristocaulodon Koch (Sichelzahn). This illustration of the *Missourium* as originally reconstructed is taken from plate 8 of *Die Riesenthiere der Urwelt,* by Dr. Koch, published in Berlin in 1845.—Reproduced from the collections of the Library of Congress

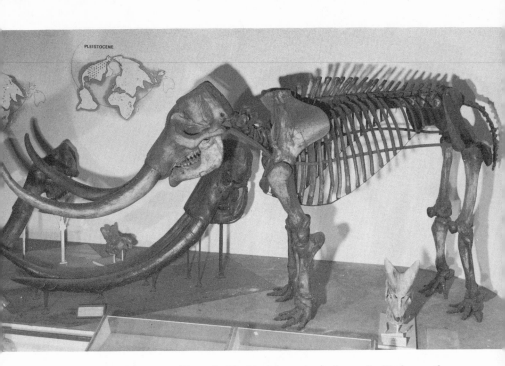

Mammut americanum (Cuvier), (Koch's *Missourium*), dug up by Koch near the Pomme de Terre River in Benton County, Missouri, 1840, and now on display at the British Museum (Natural History), London.—By permission of the Trustees of the British Museum (Natural History)

Gravestone of Dr. Albert C. Koch in the hilltop cemetery, Golconda, Illinois.—
Photograph by Ernst A. Stadler, 1969

Reise

durch

einen Theil der

Vereinigten Staaten von Nordamerika

in den Jahren 1844 bis 1846,

von

Dr. Albert C. Koch.

Nebst 2 Tafeln Abbildungen.

Dresden und Leipzig,
Arnoldische Buchhandlung.

1 8 4 7.

View of the City of Hartford, Conn., From the River—Showing an Old-
Fashioned Steamboat. Koch passed through here in his summer travels in 1844.
This illustration is taken from *Gleason's Pictorial Drawing-Room Companion,*
Vol. 4, April 23, 1852.—Courtesy Missouri Historical Society

Lighthouse at Gay Head, Chilmark, Martha's Vineyard. Koch explored this area in the summer of 1844 before heading west. This illustration is taken from the *Historical Collections of Massachusetts*, by John Warner Barber, 1844.— Courtesy Missouri Historical Society

Upper aqueduct : Western canal

Upper Aqueduct—Western Canal. Koch's narrative contains several generally unflattering accounts of canal-boat travel. Watercolor sketch by Vitale M. Garesche, August 1829.—Courtesy Missouri Historical Society

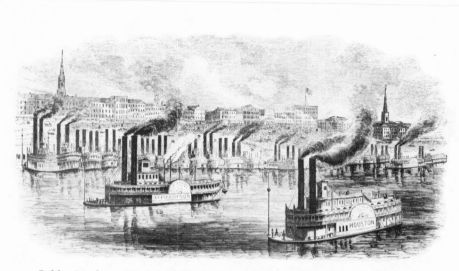

Public Landing, Cincinnati. Koch arrived here on September 8, 1844, and stopped briefly before continuing west. The illustration is from *Ballou's Pictorial Drawing-Room Companion*, Vol. 8, April 28, 1855.—Courtesy Missouri Historical Society

Cave-in-Rock, on the Ohio River. As he traveled down the Ohio, Koch, though he makes no mention of fact, passed this landmark. The illustration is from *Ballou's Pictorial Drawing-Room Companion*, Vol. 11, December 13, 1856.—Courtesy Missouri Historical Society

Map of the Ohio River and canal showing the area around Louisville, Kentucky, and Jeffersonville and Clarksville in Indiana. The original legend, in the handwriting of Meriwether Lewis Clark, eldest son of William Clark and himself an expert draftsman, describes this illustration as "from the map of the city of Louisville by Mr. E. D. Hobbs; except the plans of Clarksville & Jeffersonville & the meanders of the shore in front of them taken from a map of Survey by

Wm. Clark & a map of Jeffersonville at the office in that place." Other entries on the map are also in M. L. Clark's hand, so it would appear that the entire map is his work. Koch searched quite fruitfully for "petrifactions" in this area in September, 1844, and also observed and commented on the presidential campaign of that year. The map, undated, is from the William Clark Papers of the Missouri Historical Society.—Courtesy Missouri Historical Society

Nauvoo, From the Mississippi, Looking Down the River. Koch stopped here on his return from Bloomington, in the Iowa Territory, to St. Louis in November, 1844, and examined with interest the Mormon temple. The illustration is from *Gleason's Pictorial Drawing-Room Companion*, Vol. 7, July 22, 1854.—Courtesy Missouri Historical Society

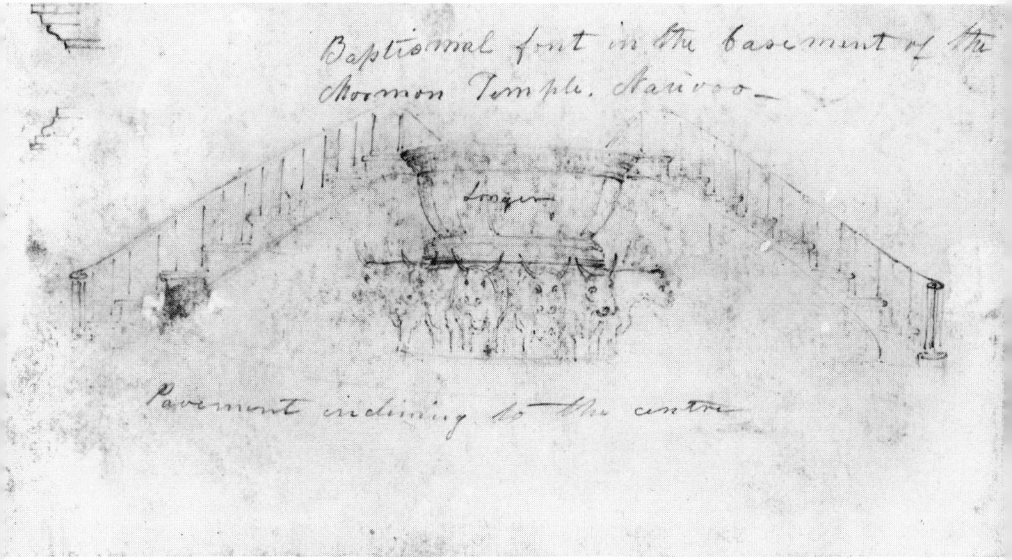

Handwritten annotations on the sketch:
Baptismal font in the basement of the Mormon Temple. Nauvoo—
Longer—
Pavement inclining to the centre

Baptismal font in the basement of the Mormon temple, Nauvoo. Koch provides a detailed description of this in his account of his journey from St. Louis to Bloomington, Iowa Territory, and back. This drawing is from the sketchbook of Henry Lewis, 1848.—Courtesy Missouri Historical Society

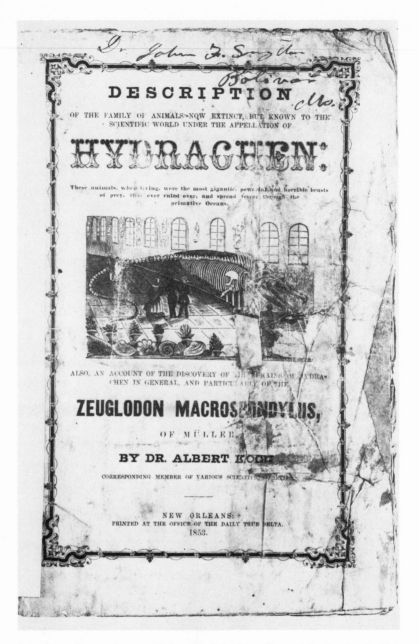

In order to finance his expeditions, Koch had to sell or exhibit his fossils. This
is the front cover of his 1853 pamphlet, *Description of the Family of Animals
Now Extinct But Known to the Scientific World Under the Appellation of
Hydrachen*, printed in New Orleans by the *Daily True Delta.*—Courtesy Missouri Historical Society

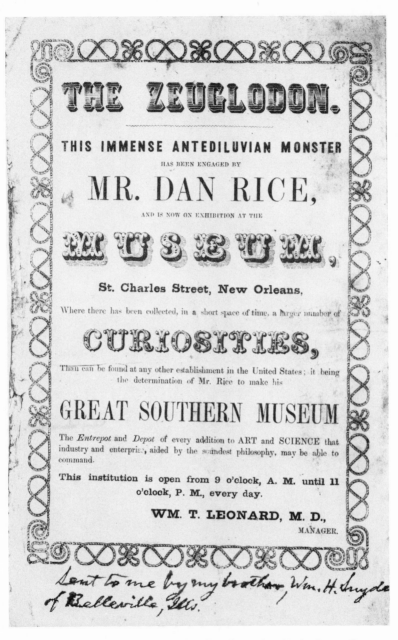

THE ZEUGLODON.

THIS IMMENSE ANTEDILUVIAN MONSTER

HAS BEEN ENGAGED BY

MR. DAN RICE,

AND IS NOW ON EXHIBITION AT THE

MUSEUM,

St. Charles Street, New Orleans,

Where there has been collected, in a short space of time, a larger number of

CURIOSITIES,

Than can be found at any other establishment in the United States; it being the determination of Mr. Rice to make his

GREAT SOUTHERN MUSEUM

The *Entrepot* and *Depot* of every addition to ART and SCIENCE that industry and enterprise, aided by the soundest philosophy, may be able to command.

This institution is open from 9 o'clock, A. M. until 11 o'clock, P. M., every day.

WM. T. LEONARD, M. D.,
MANAGER.

Sent to me by my brother, Wm. H. Snyder of Belleville, Ills.

Back cover of Koch's 1853 pamphlet describing the *Hydrachen.*—Courtesy Missouri Historical Society

These three vignettes are details from the *Panorama of the Monumental Grandeur of the Mississippi Valley*, 7½' × 348', painted in 1850 by John J. Egan. Dr. Montroville Wilson Dickeson was the narrator and exhibitor of the panorama, which was last shown in 1949 at the City Art Museum of Saint Louis and is now among the collections of that institution. The scenes depicted show workmen digging at the bed of a stream near Kimmswick, Missouri; the remains of a mastodon; and Dr. Albert C. Koch explaining the fossil remains to Dr. Dickeson.—Courtesy City Art Museum of Saint Louis

View of St. Louis, Mo. This is from a lithograph by Julius Hutawa, 1845. Koch and his family lived here.—Courtesy Missouri Historical Society

Explosion of the *Lucy Walker*. In his account of his journey from Louisville to St. Louis, Koch mentions the recent disaster wherein "106 persons on board had lost their lives." The illustration of this most horrible of steamboat castastrophes, which occurred just below New Albany, Indiana, on October 23, 1844, is from *Lloyd's Steamboat Directory, and Disasters on the Western Waters,* 1856.—Courtesy Missouri Historical Society

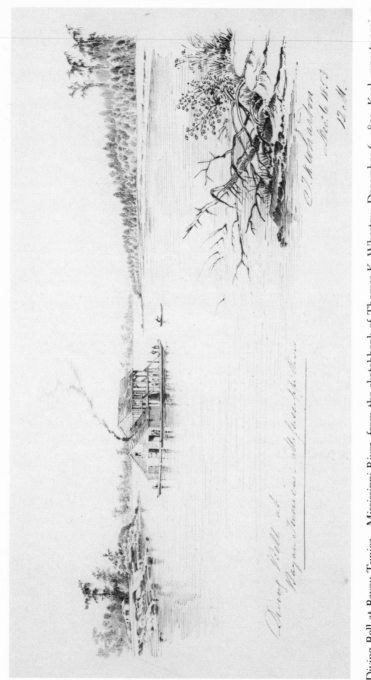

Diving Bell at Bayou Tunica—Mississippi River, from the sketchbook of Thomas K. Wharton, December 6, 1853. Koch reports seeing such a device in operation on the Mississippi on his return trip from Alabama.—Courtesy Manuscript Division, The New York Public Library, Astor, Lenox, and Tilden Foundations

View of Cairo, Junction of the Ohio and Mississippi Rivers. Koch comments on the contemporary economic problems of the town in that part of his narrative covering his trip from St. Louis to Bloomington, Iowa Territory, and back. The illustration is from *Ballou's Pictorial Drawing-Room Companion*, Vol. 11, August 30, 1856.—Courtesy Missouri Historical Society

Cotton Loading. This is the steamboat landing at Claiborne, on the Alabama River, with the staircase and cotton slide which Koch describes in detail in the section of his narrative covering Claiborne, Alabama. This illustration is taken from *Ballou's Pictorial Drawing-Room Companion*, Vol. 8, April 21, 1855.—Courtesy Missouri Historical Society

The *Hydrarchos,* as shown in the German edition of Dr. Albert C. Koch's diary. This is the sea serpent the author found in Alabama.

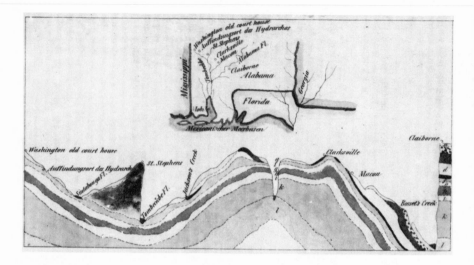

Map made by Dr. Albert C. Koch, showing geological formations in Alabama. This was drawn to accompany the German edition of the author's diary. The following description is a translation of the author's original text. (In the original, the author spells Tombigbee River as Tombeckbe.)

The appended map has the purpose of illustrating for the respected reader the geological and geognostic situation of that part of Alabama in which the *Hydrarchos* was found. The upper section depicts the southern part of Alabama and its borders, whereas the lower one gives a profile of the region of this state which offers the greatest geognostic interest, an area which extends from the city of Claiborne, lying directly at the 200-foot-high, almost perpendicular bank of the Alabama, to Washington-Old-Courthouse.

It is especially this stretch of land of approximately 70 English miles which offers the most distinct traces of earlier earth upheavals, whereby a very uneven surface, consisting of steep hills and innumerable more-or-less deep canyons, was formed. The various strata, containing organic remains, are here partly unearthed but also partly covered by a layer which does not contain such remains. On the steep banks of the Alabama, the strata are found lying horizontally, one upon another, and are marked on the map under Claiborne from *a* to *l*.

If one looks at the section from Claiborne to the place where the *Hydrarchos* was found and compares the existing sand, marl, or lime strata with those in Claiborne, one can easily see with his own eyes to which geognostic period each stratum belongs. If one comes for example from Claiborne to the opposite bank of the Alabama, one steps for a time on an upper sand stratum, marked at Claiborne with an *a*, ascending farther to the gravel stratum marked *b*. Soon after, however, the ferruginous clay stratum attracts one's attention through its partly blood-red spots; this is marked at Claiborne as the third stratum with *c*. Furthermore, after that the sandy soil *a* reappears; wherever it appears, the ground is covered with pine forest, while the remaining strata, when forming the surface, permit only the growth of deciduous trees. Near the town of Macon, which itself lies on a small plain covered with the sand layer *a*, appear intermittent traces of a white lime layer, marked at Claiborne with *d*, and the brown iron ore layer mixed with volcanic matter marked *e*. It is this white lime stratum which, as mentioned in the book, is sawed in pieces and used for building. At Clarksville one comes to the so-called prairies, which differ substantially from those in Illinois and Missouri, in that the latter consist of more-or-less vast fertile plains, while the former are areas without humus, which for the most part contain the lime substance marked at Claiborne with *f*. This stratum distinguishes itself especially because only in it are the remains of the *Hydrarchos* to be found.

A few English miles to the right of Clarksville is visible a deep canyon which offers the scholar a rare opportunity to see all the deeper-lying strata noticeable in Claiborne again laid bare. So the upper edge of the canyon shows, under the sand layer *a*, grown over with pines, the stratum *f*; however the strata *g*, *h*, and *i* are here not quite so isolated as at Claiborne, and the yellow sand here takes on a light green color. Farther down one encounters the limestone stratum marked at Claiborne with *k*, which could be called a fossil oyster bed, and below which is the stratum of green sand marked *l*. From the aforementioned canyon to the bank of the Tombigbee, the area is very similar to the one between Macon and Clarksville. Approximately 2 English miles from St. Stephens appears again the sand *a*, grown with pines, which from there to the banks of the Tombigbee, also covers the remaining strata. As soon as one has crossed this river by ferry and stepped on the ground, the chalk-white lime *d* at once appears again predominantly, and even forms here a steep cliff of 130-foot height, behind which, surrounded completely by canyons, lies the little town of St. Stephens. Intermittently, small patches of cultivated land and forest appear here, quite like the ones between Macon and the latter place. Because of the small scale of the map these patches could not be shown. Up to the discovery place of the *Hydrarchos* the region remains very similar to that last mentioned, and because the place where the *Hydrarchos* was lying has already been described in the book, it needs no further description.

Author's Foreword

W H E N I herewith publish excerpts from my diary of the journey which I undertook in the years 1844 to 1846 in the cause of widening the knowledge of the geognosy of North America, it cannot, indeed, be my intention to offer the esteemed reader much that is new, particularly since there exist already so many estimable travel books about this interesting country. I had, however, the good fortune during my stay to make some not unimportant and interest-provoking discoveries for the science of geognosy, and I therefore allow myself the hope that the following simple description of my travel adventures will meet with some approval. Hence, I hope that this little book may find the same indulgent reception I had the good luck to enjoy in my homeland as well as on foreign soil.

Berlin
March 1, 1847

Dr. A. C. Koch

From Hamburg to New York

My JOURNEY from Dresden via Leipzig and Berlin to Hamburg I pass over here with silence. Arrived in Hamburg on May 23, 1844, I let it be one of my first concerns to look at the ship on which I intended to travel to New York; it was the *Howard* under Captain Paulson. To my regret, I found not much to be pleased about because it was a ship of only approximately 250 tons, and 28 cabin passengers and 104 steerage passengers were to make the journey. Still, I came at the right time to select the best sleeping place and mark it with my name because the cabin passengers who had already marked their places had done so at random, while I knew from earlier voyages where I was best taken care of. I boarded the ship on the evening of the twenty-fifth because it was supposed to sail early on the twenty-sixth. Shortly after sunrise on the twenty-sixth we departed, pulled by a steamboat on which a chorus let its harmonies ring out, drowning out the wails of those who cried from the shore their farewell to departing relatives, as well as the sobbing of those who responded from the ship, while from both sides handkerchiefs and hats were waved as long as they could see one another. A party of close relatives of our cabin passengers went with us until the steamboat left us, and here took place the last touching farewell scene because our escorts had to go back with it. For a short time the wind was unfavorable, and we had to tack up the Elbe, but soon the scene changed and we rode with full sails, and toward evening anchors were cast near Cuxhaven; these were weighed again by daybreak to a very favorable and strong wind. The pilot was still with us, and large buoys fastened to chains showed us on both sides the navigable channel. Only on the left side, in the far distance, could we see the coast of Friesland; on the right, all land had disappeared, the waves became bigger and bigger, many faces showed considerable paleness, and ever smaller became the number who stayed on deck. Also, the cabin boys were beginning to come up from the cabin with crockery which contained indubitable evidence that seasickness had seriously started. Because I am not always the last one attacked by seasickness when it assumes a somewhat serious character, despite my many sea voyages, it is not surprising that I too became very quiet and serious, and could I have looked at myself in a mirror, I would have noticed some pallor; but I held out until we came to the last so-called light ship, named *Elbe*, where our pilot was picked up by a small vessel. (Light ships are small boats which are anchored, and which, like the already mentioned buoys, indicate the navigable channel, but with the difference that there are people on these ships who have to keep a certain number of lamps burning during the night to serve the passing ships as lighthouses, which cannot be placed here.) But now I also withdrew to my bed in the cabin; here, however, there was not much pleasing to see or hear because one's eye saw only pitiable faces, and the ear heard only groans, sighs, and still other noises connected with seasickness, which were interrupted now and then by the shout: "Fritz!" or "Peter, here is a pot to be taken away!" (Fritz and Peter were the names of our two anything but clumsy cabin boys.) I had provided myself with remedies against seasickness, but I am now more than ever convinced that there is no remedy for this horrible illness; the reason for this seems to be mainly that this evil is caused by a nerve, namely the *Nervus vagus*, which has its origin in the brain and stretches from there down into the stomach, and which by this means brings about an intimate connection of head and stomach. First comes dizziness in the head and nausea in the stomach to a larger or smaller degree, and, to the extent that this increases, it results in an emptying of the stomach, which according to the diversity of the constitution is more or less violent. Combined with this is an exhaustion of the mind, so that one is hardly able to form a clear thought and always finds oneself as if in a disquieting dream.

After a few hours we were already opposite the Island of Heligoland, which I looked at for a few minutes, but I then went again, very seasick, to my bed from which I did not rise again until we were out of the North Sea and could see the coast of England, partly veiled in fog. The wind had been so favorable to us that we, like a steamboat, made the voyage over the North Sea in twice twenty-four hours. I and many others of our seasick were again able to hold out on deck, but fog with intermittent rain allowed us to see only a little of England; still the wind stayed favorable and blew so briskly that we covered the ninety German miles, which is the length of the Channel, in less than twice twenty-four hours. Naturally we were nothing less than pleased to have survived this dangerous part of our journey because one can truly say that thousands of ships have found their destruction here.

After we had reached the ocean, the wind rose somewhat stronger but still very favorable for us; however, because of the big rolling waves, we were all forced anew to go back to our hard beds of misery. Not until the third and fourth day we were on the ocean were we able to stay a few hours on deck. Thus the tenth day of our voyage dawned. The wind had been so advantageous to us and had blown so briskly that, the captain assured us, never on all his trips had he covered such a great distance in so short a time as we had done up to now. Already we were on the same degree of latitude with the Azores and had therefore almost the smaller half of our journey behind us. Many of our passengers, accordingly, became so high-spirited that in jest they made the suggestion of drawing lots to determine which of us was the "lucky bird" who was responsible for this fast voyage. But soon the high spirits subsided and came to an end in sadness and horror. But before I turn to the now-beginning new chapter of our voyage, I must yet mention a small occurrence which happened on the second day of our trip on the ocean. Namely, we noticed in the distance a ship which tried to approach us in an unfavorable wind. When it finally accomplished this, it signaled that it wanted to communicate with us, and, as we greeted each other by raising flags and were close enough to converse through a megaphone, it turned out that it was a Swedish ship which had come from Gibraltar. Because it had constantly had to fight headwinds and was a very inferior sailing ship, it had been on the way already for eight weeks and had enough drinking water and food for only six more days. The main reason they wanted to talk to us, though, was that they did not know for certain on which degree of latitude they were and what was the nearest port where they could go to restore themselves. Our captain gave them directions, but, as they did not ask him for water and food, he did not offer them any because, as he said, we ourselves did not know what was ahead of us. But had they asked him he would surely have given them some of our provisions. So we parted without ever later hearing more of those poor people.

On the tenth day of our voyage a new phase began for us. The brisk wind, up to now so propitious for us, abated all of a sudden, and the now-prevailing deep stillness was interrupted only by the most unpleasant sound caused by the loose-hanging sails. Although no wind was blowing the sea still moved very much, and the ship, which now received no more support from the sails, was thrown violently from one side to the other, and we of course shared the same fate as the ship. This lasted, however, only until sundown of June the fifth, and this sunset already bode ill for the next days because the sun sank like a big fireball in a jet-black dam of threatening clouds, and soon a southwest wind came up, which, accompanied by heavy rain, changed during the night to a storm, and so broke the unforgettable, for me, sixth of June. I had not slept all night and was very glad when finally the morning dawned. With much difficulty I climbed up to the deck to find some consolation, but here the prospects were very disheartening. Only two small so-called storm sails were set so that the ship could be steered with them. The roar of the terribly agitated ocean vied with the howling of the storm, which raged with full fury through the masts and the rigging; with this harmonized the melancholy cries of the sailors, who struggled to make the already-mentioned sails still smaller because the storm seemed to increase rather than decrease. But all this was drowned out in irregular intervals by the waves crashing over and against the ship, which sounded like muffled cannon shots. The captain, the two mates, and some of the sailors wore southwesters and clothes made from material that was water repellent; behind the ship was a flock of stormy petrels, which, like small black spirits half walking and half flying, hovered over the waves. The whole horizon was covered with heavy threatening clouds, which were whipped away by the storm to make place for some new, possibly even more terrifying looking clouds, some of which almost touched the waves. Because this could fill me with nothing but feelings of dread, I asked the captain whether this gruesome scene could possibly be of any duration, and received the answer that if it were fall or winter he would have to believe from all the signs that we would have to weather a storm lasting eight days; he hoped, because it was June, we would come off a bit easier, but the barometer held out small hope. After this answer I moved awkwardly toward the barometer to see for myself the sad reality of our chances, and here, unfortunately, I found the matter even worse than the captain had told me because the barometer predicted almost an earthquake. I had seen enough to prepare myself for the worst, and I climbed and crawled back down to my bed, commending myself in fervent prayer to my heavenly father and guardian and supplicating him for his special help in our distressed situation since, as our prospects were, it looked now as if human help could not do much. Only through much effort and precaution was I able to stay in bed; most of my fellow sufferers were already lying on the floor amid trunks and so on. The storm became more violent, and by ten o'clock it had become a terrible, all-destroying hurricane. The whole sea was now continually covered with white foam, and each single wave seemed like a mountain with a whirling eddy at its crest. The ship lay so far to the side that the yards touched the water; the motion was frightful, and the whole hull trembled so that one might think that all the joints would come loose. At about half-past ten the ship made a horrible shuddering motion which was accompanied by a rumbling noise lasting a quarter of a minute but which was different from the one caused by the waves going against or over the ship. At the same moment, one of our companions, with a deadly pale face on which showed an expression of desperation and hopelessness, jumped up from his place on the floor and cried: "We are all irretrievably lost!" Whoever has enough imagination to place himself in our situation can imagine the impression these few words made on us—and, in addition, the wails of lamentation and the weeping of the women and children (we had ten women and girls and five children in the cabin). As already noted, we had had until this time two small lugsails up, but now even those were too much, and the force of the hurricane threatened to sink the ship. The sailors were just trying, under the greatest danger to their lives, to free the ship from this burden, when one of the most violent thrusts of the hurricane decided our fate. The power of the storm was now so great that we must sink if the sails did not immediately come down. But God had merciful pity on us poor helpless creatures, for in an instant both sails and part of one mast fell overboard. Yet here again it was proved how remarkably at times dangers are crowded into one moment and how our heavenly protector visibly stands helping at our side when human help is impossible; in the just-mentioned horrible moment when our sails tore, a large ship had been driven so close to us that only a part of a single wave separated us from it, and everything depended on whether this wave surged away from us or toward us, for in the latter case both ships were past saving. But this wave powerfully seized our dangerous neighbor, and soon he disappeared from our view and we were saved from twofold great danger because, now that our ship was freed from its sails, it became considerably lighter and had lifted itself a great deal. Besides, the hurricane was too strong to last any longer, and soon it changed again to a storm which lasted with great fury for three more days and three more nights. After that followed a stillness of the wind similar to the one already described, which ended after twenty-four hours with a new storm. It would become boring if I were to describe the subsequent storms and lulls, enough of that; in 15 days we progressed hardly five degrees and were exposed to many dangers. Then the wind changed a bit in our favor, but we still were thrown from North to South and back to North, so that toward the end of our journey we had come at one time within ten or twelve miles of the Newfoundland coast, and later found ourselves in the Gulf of Mexico [Gulf Stream?]. Several times we had a chance to observe whales very close to the ship and to see the spout from their air holes, and I believe if we had been on a whaling expedition we would have done a good business. I also found proved on this voyage an earlier observation, that, when the so-called dolphins appear in masses on the surface of the ocean and jump in curves out of the water, twenty-four hours later either rain or a storm commences, or both together. It cannot with certainty be stated why just at that time these animals like to keep themselves on the surface of the water and jump; but I am of the opinion that at that time movements take place on the floor of the ocean which cause those creatures to amuse themselves above.

During the last-mentioned phase of our voyage, the sailors and mates had to work so hard and were day and night so completely drenched not only by rain but also by the waves washing over the deck that they developed big sores on their hands and arms, and four of our seamen, as well as a mate, became totally useless. Unfortunately we had in all only ten sailors and two mates; if the incessant, one-after-another storms had not soon abated a bit, God knows what would have become of us. The weather was even now scarcely fair and the wind barely favorable, but our wishes in this respect had become so modest that we counted ourselves lucky when we could find a spot on deck where we were comparatively safe from the rain and the waves which sometimes washed on deck, and when the cold had abated to the point where one could stand it on deck with just a cloth jacket or coat, which surely is a modest demand for the middle and the end of the month of June, since one should, rather, wear summer clothes at this time; but I spent only the last two days of our journey on deck in a cloth jacket. As pleasant as some warmer weather would have been for us it could also have thrown us again into great distress, because in all probability the bad air would have caused illnesses, and our drinking water supply would not have lasted, as the ship was almost overcrowded with passengers of all classes.

As I already remarked, the movement of the ship was always such that one could not think of getting exercise by walking to and fro. Indeed, it was usually difficult to get from the cabin to a seat on deck, and I believe there was no one among our passengers who did not one or more times have a rough fall; some even suffered minor or serious injuries. I myself, among other things, once had my foot slip at a violent movement of the ship and fell down the stairs, whereby I injured the right side of my chest, causing me much pain which I slowly got rid of by applying a mustard plaster.

Until the eighth of July nothing noteworthy happened on our voyage; but when I awoke early that day I felt by the movement of the ship that the wind had changed in our favor, a fact which to my joy I found confirmed when I came on deck; all day long the wind was favorable to us and blew briskly so that we could soon look forward to our final deliverance from the ship. By daybreak of the ninth I was again on deck to see how we were making out; we had made considerable progress during the night and were now fully under way. In the afternoon the captain began to look for a plumb line, and after seven o'clock the cry "Land!" sounded from various sides, and it was truly touching to see the impression which this word and the sight of land, looking like a foggy cloud in the distance, made on all. Several of us broke into a stream of tears of joy and thanks and pressed one another's hands. Others tried to hide their emotion but could not keep some tears from rolling down their cheeks, and I am not ashamed to confess that with wet eyes I sent a silent prayer of thanks up to Him who had so visibly saved us from the manifest dangers of death, and only he who has had a similar experience can form a true notion of such moments.

As soon as it became dark we saw first one and then several light-houses, and a lantern was hoisted on our ship to let a pilot cutter know that we needed their help. To our great joy our signal was answered toward ten o'clock, as we saw in the darkness a light approaching us with great speed, and shortly an extremely delicately built small boat floated around us like an agile water bird. The question of whether we needed its help was at once answered in the affirmative. Because our sails were now all taken in, or at least so fixed that the ship could not get closer to land, the pilot's first concern was to get it going again after the plummet had been transferred from the pilot boat to our ship. Shortly we moved up in the mouth of the New York harbor and dropped anchor for the night because it was not thought advisable to go on now. By daybreak the journey slowly continued, because only a light wind blew which, however, was favorable to us. Everything now was joy and anticipation, and we took delight in the sight of the lovely banks of the harbor until we arrived at about ten o'clock at Staten Island, where we were put ashore and the ship had to stay in quarantine for two days. Here, after having endured so many dangers, one congratulated another with a handshake on the safe arrival on American soil. We then went by steamboat to New York, where we scattered in all directions. My Dresden traveling companion, Herr Advocat Ludewig, his wife, a young fur merchant from Luebeck, and I climbed into an omnibus together. On our voyage we were so unaccustomed to the warm weather and to walking that, although we had only to walk a few steps from the spot where the omnibus deposited us to the house where we were staying, we arrived there exhausted and soaked with perspiration.

Because all those whom I wanted to see in New York were out in the country, I did not stay in this city any longer than circumstances required. I visited the local American Museum, a fairly good collection of natural history objects where there are, nonetheless, only a few fossils. Among those few, however, I made the discovery that we in America have, besides the hitherto-known mammoth or antediluvian elephant, still another species of the primeval elephant, unknown until now, which has almost as great a similarity to the now still-existing African elephant as the mammoth has to the still-living Asian one. In the mentioned museum there are three molars of that animal which were found in Kentucky and are thought to be, out of ignorance, mammoth teeth. When I shall have had the opportunity to examine these teeth more closely I shall say more about them.

Since I hear that several ships which were destined from Liverpool to New York are missing, I must note that we saw at four different times sealed bottles floating close by our ship, which without doubt came from shipwrecked seafarers and contained news of their terrible fate. Warships have the obligation to pick up such bottles, but merchant ships can act at their own discretion. One day we even encountered single pieces of a ship.

✤ New Haven, Hartford, Springfield, Boston, New Bedford

ON JULY 18, at six in the morning, I left New York on the beautiful new steamboat *Champion* to travel to New Haven. It was a glorious summer morning, and, because it had rained very hard two days earlier after a long drought, everything looked newly refreshed. One goes up the Long Island Sound and has on the right the loveliest view of fine country houses, built in varied styles, and the gardens and fields of Long Island. To the left can be seen the shores of the mainland, where one's eye sees soon a friendly little town and then, standing up on cliffs surrounded by green pines and deciduous trees, some white-painted lighthouses which by night warn the traveler of threatening danger with their brilliant light. Yes, one can truly say the eye is offered so much beauty and variety that one doesn't know where he should let his gaze linger.

At eleven o'clock we arrived in the harbor of the city of New Haven. Far away in the distance to the right can be seen like blue fog-clouds the shores of the last tips of Long Island. To the left opens the really romantic, though small, harbor of the city, which is enclosed all around by rocky, primitive mountains. The slopes and the crevices where soil has formed are covered with wild grapes, flowers, and trees. But these mountains are far enough away from the harbor so that there is space for an almost level valley in which the city of New Haven lies like a paradise inhabited by people. It may seem presumptuous or exaggerated that I use the word "paradise," but if there is one city and its surroundings, inhabited by people, that can be compared to paradise, it surely is this. With the exception of a few crammed-together business streets, the whole city of New Haven, which is 80 English miles away from New York, looks like one of the most beautiful parks. It is cut through with big, broad, shady avenues with wide and clean sidewalks for pedestrians. Each house is set back a bit from the street in a tastefully laid-out garden of English style which is surrounded by a beautiful wrought iron fence. A wide and straight walk leads from the street to the house, which is decorated with several columns, has a flat roof, and proclaims taste, beauty, and affluence, if not wealth. Although the city counts only 11,000 inhabitants, it stretches over a considerable area. In the middle is a large square, which is not paved as our market places are, but is surrounded by an avenue with very large shady trees and crossed by walks which connect to the various streets. Moreover, this whole big square is planted with grass on which stand high, broad-limbed trees. In short, everything seems to be calculated to make the stay as comfortable as possible for those who live here as well as for those who just pass through. Even with the place of final rest the citizens of New Haven have taken great care. They have a very large and beautiful cemetery directly adjoining the city. It is surrounded by a granite wall and has a painstakingly worked entrance. Each family has its own burial place, and a considerable space has been reserved for the poor.[1] The city possesses a university [2] with 500 to 600 students, at which the famous Professor Silliman, in whose family circle I spent most of the time of my short stay in New Haven, has worked with the best of results for 40 years.[3]

On the 19th of July at eleven o'clock I rode on the train to Hartford where I arrived about two o'clock. The passenger cars here are quite different from the ones in Europe. Here, each of them is a long, narrow room with continuous sliding windows on both sides and an aisle in the middle; on each side, right and left, are two very comfortable armchairs, so that in each of these cars 48 persons can sit very comfortably. Hartford lies on the Connecticut and has more commerce than New Haven, but it cannot match the other in beauty.

At half-past two I went on a small steamboat called the *Agawam* from

Hartford up the Connecticut to Springfield, a smaller but far nicer city than Hartford. We arrived here after seven o'clock in the evening and had to spend the night here. The accommodations were very good, and, to give my German readers an idea of an American evening meal and breakfast, I only want to say briefly that we had warm food both times. On the elegantly decorated table we found coffee and two kinds of tea, two kinds each of warm and cold bread, fried sausage, beefsteaks, fried chicken, fried ham, fried doves, eggs on butter and soft-boiled eggs, black raspberries and blueberries (from which one does not get a black mouth), two kinds of *torte,* several kinds of cake, and, finally, cheese and butter.

On the 20th of July, soon after 6:30 in the morning, I rode by train to Boston, where I arrived at twelve noon and took lodging in the United States Hotel, an inn which can house 400 to 500 guests with the utmost comfort and which has the convenience of being placed right across from the railway station. It had rained all day, and the streets were therefore very muddy, but the coming day was a Sunday, a day on which you cannot do any business in America, least of all in Boston. Therefore I hurried to deliver my introductions, but I found only a few of the ones to whom they were directed at home. Nevertheless, among the few I met was Mrs. H. G. Otis,[4] a lady of distinction and great influence in Boston. I was received by her with great courtesy and kindness and could only regret that she could not introduce me to any of the learned gentlemen of Boston because almost all of them were out of town.

Though Boston has much to please, I had on the whole imagined the city to be much prettier. The nicest place here is the so-called Commons, a spot with the circumference of an English mile, which is bordered on three sides by some of the most splendid houses of the city and on the fourth side by water. A nice high walk circles the Commons; the whole spot is covered with a splendid lawn, shady trees, and many intersecting walkways.

I had intended to go to Church Sunday morning, but, when I came home the night before, I found a card informing me that a Mr. Simmons, whose brother I had met in Berlin and who had visited me in Dresden before my departure, wanted to visit me the next morning. I found him, as I had his brother, to be a very pleasant young man. He took a walk with me this morning to Charlestown, that famous place of the American wars of liberation, where I wanted to see the newly finished monument on Bunker Hill.

This monument is a masterpiece of modern architecture; it consists of an obelisk built of square-hewn stones of gray granite, measuring 120 feet in circumference and 240 feet in height above the ground. The whole thing is massive, with the exception of a spiral staircase in the middle which leads all the way to the extreme tip, where there are four vistas which must afford a magnificent view, which I could not enjoy because it was Sunday, when almost everything which would give diversion and secular pleasure is closed up in Boston. The cornerstone was laid by General Lafayette many years ago.[5] The foundation alone cost several thousand dollars, and, because of the large expense, the construction was halted for several years and was finished only a year ago. The greatest difficulty lay in the transporting of the immense capstone to the top, and there were more than 500,000 people assembled to see the stone put into place.

Because much has been written about Boston, and because my stay there lasted not quite two days, I shall not make any more remarks about it.

Through Mrs. Otis I made the acquaintance of a very cultured Swiss named Mallignon, who, like me, is staying at the United States Hotel, and who arrived only three weeks ago to work as a French teacher. Because he was warmly recommended to the first families of Boston, and receives three to four dollars for each lesson, I am convinced that he will make his fortune.

Monday, July 22, at eight o'clock in the morning, I departed from Boston by train to go via New Bedford to Martha's Vineyard. Shortly after Boston I noticed, on the road to my left, a rock formation of great geological interest. Although the train went too fast to obtain a better idea of that formation, I nevertheless saw that it was a very special conglomerate which seemed to be of secondary formation. I decided to stop on my way back and conduct closer examinations. Until New Bedford, which is 56 English miles from Boston, I found nothing noteworthy. I arrived at 10:30, and right away had my things brought to the harbor into a small, dainty, four-masted ship which connects this place and the town Holmes Hole,[6] on Martha's Vineyard. Although a steamboat also travels the same route, it runs only three times a week and was expected in only today. New Bedford, as is generally known, is a place from which many whalers set out. Two of these ships, of which one was away for two and the other for four years, had just arrived; I heard that both were very satisfied with their catch. One of these ships carried 3,200 barrels of whale oil and had therefore caught approximately 32 whales; how many the other ship had caught I could not find out. At eleven o'clock our boat set sail; the crew was very small, but the peculiar thing was that there was no common sailor among them. It was run by two very deft men, of whom the older was introduced to me as the captain and the younger one as the mate. Besides me, there were three other passengers on the boat, of whom one was the captain of a similar vessel, which are known here as packets; the second was a German merchant from Nuernberg who had lived in America for seven years and who seemed to be very happy here, and the third, a young man who lived on a neighboring island. The wind was rather favorable, and so we covered the 28 English miles to Holmes Hole in four hours, landing at three o'clock. I had fun watching the captain and the mate alternate at the helm, and when both were needed to change the sails, how they fastened the rudder with a rope and left it alone. It must take quite a lot of experience and deftness to approach the harbor of New Bedford, because we passed many rocky cliffs, and close to Holmes Hole lay a boat which had been shipwrecked there last winter. Holmes Hole is a friendly town of 1400 inhabitants and lies on the slope of a hill which bounds the small harbor.

✤ Martha's Vineyard

TUESDAY, JULY 23rd, after ten o'clock, I started my journey to Gay Head, the lowest part of the island and the first destination of my trip. From Holmes Hole to Gay Head is a distance of 18 English miles; the first 11 miles of the way can be traveled without risk by carriage and I rode the stretch with a man who, twice a week, carries the mail in a one-horse carriage. In Holmes Hole I had already heard that the lower part of the island was a United States reserve, which was not especially apparent to me, since a lighthouse is located here; but to my not small surprise I learned also that part of Martha's Vineyard is to this day inhabited by Indians, who make up the last remnant of the Gay Heads, a fact of which I had never before read or heard.[1]

The part of the island which I now had to traverse on foot was very barren and showed traces of great earth upheavals which had taken place in prehistoric times; it is therefore no wonder that our white fellow citizens left this desolate region as a last refuge to the poor Indians. I left the carriage at the last house, which was surrounded by a few trees. From here the whole area showed only bare hills divided by somewhat more fertile valleys which were frequently broken by ocean inlets and small sand steppes. Yet those hills and desolate valleys were separated into irregular fields by man-made walls of field stones, and here and there rose a house, which, while giving evidence of a certain prosperity, looked lonely and melancholy indeed without a garden or the shadow of a tree. The tired traveler asks almost instinctively: "How is it possible that people who were able to build such houses and walls, at great expense and no small amount of trouble, did not turn to another friendlier and more fertile region of America, which this continent offers in such variety?" The study of this question yields the conclusion that love of the land holds them here where their ancestors had lived from time immemorial, and where they themselves had been born and had always lived. But the whites tried a few years ago to challenge their rights even to this last desolate spot until the governor of the state took up the cause of the helpless Indians, one of the oldest of the Indians told me.

After a very arduous march I arrived at the house of the old Indian who, I had been told, would, for pay, give lodging to strangers; at my arrival the master of the house was not yet at home, but only his wife and a young girl were there. To my question as to whether they would accommodate strangers, the brown woman replied: "Sometimes!" To my further question, whether I could stay overnight, it was said: "When my husband comes we will think it over." To my joy he did come very soon and put an end to the ceremony; I was now led from the kitchen, where I had been received, into a very decent room in which a large genuine American double bed played the leading role. It was quite well furnished in the customary manner, even to papered walls, and, in the absence of a rug, the floor was painted with a brown oil paint. I mention this only to show the degree of civilization the local Indians have reached. The food I found quite in keeping with the already mentioned furnishings, for I ate and drank better than I could have expected to.

As tired as I felt I was too eager to see the spot which had chiefly drawn me here not to visit it on the same evening. I went therefore, after I had refreshed myself with some rest and a meal, to the lighthouse, a distance of only one quarter of an English mile. I had to hand over a few lines from Professor Silliman to the lighthouse keeper requesting him to assist me in my endeavors. At first he made a long, doubtful face, but when I, noticing this, mentioned casually that I was staying with his Indian neighbor, his mien brightened noticeably and he was at once willing to show me the location of organic remains that had lured me here. In a few minutes we stood on the extreme edge of the island, at least 200 feet above the ocean surface, and had below us and to the right and the left a view such as I had never before seen and which was so special and original that it is very hard for me to give a description of it. For years the high, steep banks have been washed into the deep by the ocean surf and have entirely taken on the form of glaciers leaning against the land, but with the difference that glaciers have their beautiful play of colors only when the sun causes their rainbow colors, by the breaking light beams on them, and, further, that this color-play is only partial and weak, while, on the shores of Gay Head, all colors of the rainbow show themselves in such a brilliance and such a beautiful fusion as only the richest fantasy of a painter could imagine. As a result, the landscape takes on an almost unearthly and magical appearance which probably has no equal in the whole world. The predominant color is a dark rose-red which becomes ever softer and turns into a beautiful gray that gradually rises to the deepest brown and black and then again fades little by little into the aforementioned red, which now clears to a white, pure as that of freshly fallen snow, which sparkles magnificently in the sunshine. This turns into a light blue which again changes to a dark black or sometimes red; there is also some yellow here and there. The tops of those colorful, wildly ragged hills display low but very dense bushes mingled with many vines of Virginia creepers in the most luxuriant spring green. After I had delighted myself on the just-described view for a time, I started to study the kind of soil which produced these interesting color changes, and I found that Gay Head, in respect to its geology, too, is one of the most curious places I have ever seen. As already mentioned, the red color plays a leading role here, and so we will start with it. It results from a large mass of the best red ochre, which is found here in such quantity that in stormy weather the waves wash it off so that the ocean is dyed blood red for one English mile. In earlier times the local Indians painted their houses red with it (a color they especially love). The white comes first from an alabaster-white special sand, which I cannot remember having seen anywhere before, and which in appearance has much similarity to kitchen salt, secondly from a very beautiful white pipe clay, which is found in abundance and sold by the Indians to the whites for clay pipes. The blue originates from a blue clay which likewise is frequently found here. The brown and black result from not-insignificant veins of very good brown coal—which is not used, however, because of the total lack of wood. Peat, which is found in great quantity in the hollows of the island, is burned. The yellow is part of a great mass of fine ferruginous sand which covers one of the most remarkable conglomerates consisting of round stones, flint pebbles, and sand mixed with a large quantity of primeval shark and saurian remains. Unfortunately these are very shattered and are joined so firmly with the stones and sand that it is a most difficult task to separate them from each other. It was therefore foreseeable that it would cost me much trouble to assemble a collection of these most interesting remains, which was the object of my study here. I shall now let my observations follow in the form in which I had written them in my diary.

Thursday, the 25th of July. Yesterday I worked very hard and became more and more convinced of the difficulty of my present undertaking because the vein, or thin layer, which forms the already-mentioned conglomerate is at least 150 feet above sea level and is covered with a layer of sand 20 to 50 feet thick. Like a big wall creeper which picks out insects from old masonry, one has to place oneself half-sitting, half-lying, and with iron instruments probe for, and if possible extricate, the objects of value. As a helper I had taken along the old Indian with whom I now lodge, especially to become more familiar with the locality. We found 15 teeth and a few vertebrae, the latter belonging to saurians, but the majority are from two species of sharks which were distinguished by their size—some of them must have been 50 to 60 feet long and perhaps even longer. One species has teeth with sawlike edges, and the other, with sharp, smooth edges; the first species seems to have been the larger one. I saw vertebrae of saurians which could have measured 30 to 40 feet in length, but these vertebrae I could not fully dig out. A very peculiar circumstance is the fact that among these remains of ocean creatures are found not only many pieces of wood but also extremely large pieces of well preserved charcoal which are firmly cemented together with the bones and the pebbles, as is shown by a piece I have kept.

This morning I continued my research and work, and indeed with somewhat better luck than yesterday. But suddenly I felt ill, and I hurried as fast as I could to my residence. I had a severe colic, stomach ache and nausea, and only then did I remember that yesterday I had a few times drunk too much cold water because the hard work, to which I had long been unaccustomed, made me very thirsty. My indisposition abated, however, in a few hours. Today it rained all afternoon and so I have the comfort of not having missed anything.

Friday evening, the 26th of July. My collection has increased greatly today; but I am also so tired that I can hardly write. It is already half-past eight o'clock and I have just now finished sorting out, to a certain extent, what I have collected today. I was particularly pleased that the lighthouse keeper showed me a spot where I found many parts of primeval crabs. These crayfishlike creatures appear in a deeper formation than the conglomerate in which I found sharks' teeth, and that belongs, as already mentioned, to the Eocene formation. Those crabs are from a secondary formation and were deposited here by a mighty water current together with a layer of so-called upper green sand in which they are found petrified only singly and in damaged pieces. As troublesome as the business of collecting here was, I was nevertheless rather lucky in my finds, and I am very satisfied with my day's work.

Saturday, the 27th of July. This morning I started working very early and had also much luck with my collecting, though it was done with great bodily exertion. The most extraordinary thing I discovered is an incisor of a creature which to my knowledge has not yet been described, but I am of the opinion that this tooth belongs to a very large saurian.

Unfortunately I could not find the lower part, which makes classification of the animal difficult, but I hope to be able to say more later on after I have had the opportunity to make comparisons. Until then, only this much: the part of the tooth I found has a length of 4¾ inches and ¾ inches diameter; it is almost straight, and on the side, where there is a noticeable small bend, it shows somewhat more surface than on the other parts. The thinning out toward the point is very slight. The point itself gives evidence of much use of the tooth by the animal. A second very interesting piece which I discovered today is the upper and middle part of the humerus or bone of the upper arm of a saurian of colossal size. Because this bone is as big as the one of a medium whale, and because I found the incisor and two very large vertebrae together with this bone, I am almost of the opinion that they belong to one and the same animal.

Sunday, the 28th of July. Toward eleven o'clock I went, accompanied by my host, to a church service which was held in a large schoolhouse a distance of about 1¼ English miles from here. The house lies on one of the most imposing elevations of Gay Head. From three sides one has a view of the ocean; on one side the Elizabeth Islands show not far away, while one sees on the other side even closer the little island, No Man's Land,[2] on which three families live. But here also as far as the eye can see there are no trees. Their place is taken by rather frequently appearing masts with white sails of boats, small and large, which are occupied in catching fish. The swordfish, which is caught with a harpoon, seems to be the fish most pursued here; his meat is taken from here to New Bedford, where it is sold for 3 to 5 cents, or ca. 1 to 1½ silver pennies, a pound.

When we arrived at the schoolhouse, we found standing near the entrance a number of brown men, among whom the preacher stood out in his clerical garb, such a strange man that the pen of a good writer could give a most interesting description of him. He is blind, can neither write nor read, and had not the slightest opportunity to train himself for the position which he has now been holding for twenty years. In addition, this man is one of the last descendants of the local Gay Head Indians, and he plans to die here. Already in his earliest youth he contracted an eye disease which, through improper treatment, led to total blindness. He is a man of approximately 45 years, of more than medium height, and, like all local Indians, of very dark color; his naturally black straight Indian hair is beginning to be mixed with gray, his face has something engaging and friendly which I cannot remember having seen on a blind man. To make the defect of his eyes less shocking, he wears green eyeglasses during the sermon. Because he himself cannot read, he has somebody read to him from the Bible, or whatever else he wants to hear, and he is said to have a strong memory almost without parallel.[3] After we had waited a short time outside the door we went into the house, where many women were already sitting on the left side; the men occupied the right one, and shortly almost all seats were taken.

The local inhabitants are for the most part Baptists, and I heard during the sermon that a young man who is half-Indian and half-Negro would be baptized right after the service. The sermon was, to be sure, not scholarly, and also somewhat disjointed; all the same, it should astonish anyone, considering who preached the sermon. Toward one o'clock the service was over, and we just had time to eat lunch hurriedly before the commencement of the baptism, which took place very close to our house, in the ocean, on whose flat shore all were assembled singing. The preacher still wore his cassock, but now he had girded himself with a rope. The young man who was going to be baptized was lightly dressed and he wore a cloth bound around his head. After the singing was over, the preacher knelt and said a loud prayer; then he went alone up to his hips into the water; hereupon he came back and took the neophyte with him. When both stood up to their hips in the ocean, the baptism proceeded with the same words we use, but with the difference that the preacher grabbed the neophyte with one hand behind the neck, and with the other in front of his chest, and then submerged him backwards completely in the water. Then both stepped out again onto the land, and the baptism was over.

Monday, the 29th of July. Our lighthouse on Gay Head, whose splendid light I enjoy every evening, has started to sink because the ground is giving way. To remedy this defect one makes use of a method which our German officials no doubt would hardly have hit upon, and which seemed to me indeed very peculiar, but nevertheless proved practical. Several times I have seen houses being moved here from one street to another (it is especially customary in Michigan),[4] but that this could be done with a lighthouse I would never have thought possible.

But this morning the local lighthouse still stood in its old place, and this evening it is already in its new location approximately thirty steps from where it stood before; it is, however, like many American houses in small towns, made only of wood.

Tuesday, the 30th of July. To leave nothing uninvestigated during my stay on Martha's Vineyard I undertook an excursion, accompanied by a half-Indian to the part of the island which is opposite my present abode. We followed the seashore to investigate the ravines but found no formations or layers which could contain objects of interest for my paleontological collection. After we had walked for several hours we reached the house of a man to whom I had wanted to speak for a long time but whom I found only now after a long search. The reason I looked this man up follows. When I visited the famous geologist Lyell [5] for the first time in London, he showed me the head of a walrus which, he told me, he had brought back from a visit on Martha's Vineyard and which he thought to be a fossil. Mr. Lyell at the same time noted that the whole skeleton of that animal had for some years been visible on a slope of the seashore but that nobody had thought it worth the trouble to pick it up, until finally the head was taken by the man from whom he had purchased it. The appearance of this new head, however, struck me very much, for it still had an entirely fresh white color, and its bones and teeth were very well preserved. But because it was pointed out as a fossil by such a famous geologist, I did not dare contradict him and only asked whether Professor Owen had made an anatomical examination and, if this was the case, any anatomical differences from still-existing walruses were noticeable. I received the answer that Professor Owen had found some, but I was not told which. The whole thing left me very unsatisfied, and I then resolved in secret to investigate more closely if I should ever come to Martha's Vineyard. The more so because if this island so close to scholarly Boston had shown for years on its shores such paleontological treasures as the skeletons of primeval walruses which went almost unnoticed, this neglect would have to be regarded at the least as a very severe transgression against science and one of which the usually enterprising and diligent eastern Americans here would have been guilty. For all those reasons I was extremely pleased to have before me now this man who had found the aforementioned walrus head. I was told that it was found approximately four years ago on an elevated seashore two English miles from the local lighthouse. The young man who found it declared that he never saw anything of the other parts of the skeleton, and that he was the first and only one who made this find. On my way back I examined the place of discovery and saw that it consisted of a postdiluvian formation, namely a mixture of sand and clay with pebbles and larger round stones washed ashore. The shore here is about 12 to 15 feet high, and the head was found six feet above the surf. I investigated at various places but could not discover the slightest sign of other organic substances. From this it follows that the walrus head found here is not a fossil one, but it came here accidentally and belongs to still-existing species of those animals.

Because different opinions exist in regard to the formation or layer in which many organic objects are found here on Gay Head, I cannot help but make some more remarks on this subject. For it is maintained by some geognosts that the layer which contains these fossils does not belong to the Eozoic formation, as had been assumed by Professor Silliman and others, but to a younger formation. This error seems to have developed because the shells which are found here, and of which I collected a considerable number, are located in sand or gravel containing much iron which forms the upper layer that contains fossilized organic objects; but, besides three species of shells, I did not find anything in that layer, although I worked daily in it. This layer is covered by a second one which contains many organic objects such as bones and teeth of various creatures which, firmly cemented together with gravel and many round stones, form a very hard conglomerate which I have mentioned earlier. To which geological era belong the shells which I have named but which are not in the conglomerate, they themselves give evidence. Everything that is found in that conglomerate gives evidence that it belongs to the old Eozoic formation. But now comes still a third layer containing organic remains, which is older than both of the already mentioned ones, and which consists of green sand and contains bones of saurians, sharks' teeth, and the first traces of crabs and so forth. Below this layer is found a fourth one of pieces of blue clay in which are found the last organic remains which I noticed on Gay Head. In this fourth layer show many remains of various crablike creatures, single pieces of saurians, and also some sharks' teeth. Below the last described layer is located a dazzling white clay which is shipped from here for manufacturing clay pipes and other purposes. This clay is mixed with round pieces of granite but contains, as already mentioned, no organic remains.

Having now made these remarks based on practical experience, I will return again to the house where I made inquiries in respect to the aforementioned walrus head. After I learned what I wanted to know about it, I was told by the man to whom I talked that he was in possession of two of the largest shark teeth which possibly have ever been found on Gay Head. That I welcomed this news much more than anything I had heard here until that time is well understandable; but how great was my astonishment when I saw those teeth can only be felt by someone who has seen them himself. Considering that they belonged to a shark, their size seemed almost incredible, and the ones which I had so far collected are in the same proportion to these two as the teeth of still-existing sharks are to my specimens. That I did not let this treasure escape me but acquired it as soon as possible is hardly necessary to mention.

From here we went on until we reached the farthest goal of our journey, namely a hill cut through by a road, which contains a layer of the same conglomerate which is found here on Gay Head. At the time of the road work, several sharks' teeth, etcetera, were discovered, but now it would take much work, hardly worth the trouble, to collect here, especially because wages are extraordinarily high and the laborers take their work very easy (one cannot get a worker for under $1.25 a day). I soon turned back again, convinced that Gay Head was the best and most productive place for my purpose.

Thursday, the 1st of August. The continuing rain prevented me from engaging in large explorations. Not long ago I heard that the Gay Head Indians have the proper tribal name Piequatto Indians. Only two of them are still living who can speak their mother tongue; all the rest speak only English. With these two, then, the language of the Piequatto will be lost. They bury their dead in a sitting position and put all their personal effects into the grave. They seem to have lived mainly from fishing, which is still very profitable here. Because I had never formerly had a chance to eat swordfish, I was pleased to be treated this morning to a nice piece of fried swordfish by my attentive old landlady, and I can say that it tasted good beyond all expectations.

Friday, the 2nd of August. Almost daily Gay Head is now visited by tourists since it is, as I have already mentioned earlier, a most remarkable peninsula which is connected by a very narrow strip of land with the much bigger island of Martha's Vineyard, of which it is therefore counted a part. This afternoon even the Governor from Boston [6] came to admire Gay Head. Besides his wife, there were various other ladies and gentlemen in his retinue so that, with the servants, there were twenty people. Their headquarters are at the lighthouse but, because not all could stay there for the night, four gentlemen—a sea captain, two doctors, and a private citizen—will sleep with us.

Saturday, the 3rd of August. The gentlemen who slept here last night seem to have enjoyed themselves; we sat up last evening talking until after eleven o'clock, a pleasant change in contrast to my usually quiet evenings. This morning it was intended that I should have the honor of a visit by the Governor, but I had only the pleasure of meeting his wife and some other ladies and gentlemen. The Governor had to leave the party when the lighthouse keeper called for him to make some arrangements concerning the lighthouse. The whole party took great interest in my local collection and bestowed praise on me for my diligence because I had done more than anybody else who had been here before. Regarding the last remark they are not quite incorrect, although I do not want to praise myself, because if I really deserve praise I want to let my collection speak for me. After ten o'clock our illustrious visitors departed, and I started anew on my work.

From today's guests I learned that in all the western states various sorts of fevers were raging because this spring almost all their rivers had risen higher than anyone could remember, and the drying up of the residual water would of course cause much illness.

Sunday, the 4th of August. This morning I heard our blind preacher for the second time; he preached as beautiful a sermon as could be heard anywhere. It was so true, so well arranged, and altogether the text so well applied that I was not only very edified by it but also very astonished.

Monday, the 5th of August. This morning a boat arrived to take on a load of white clay, which is used in great quantities by alum factories as well as for the already-mentioned purposes. When such a ship arrives an Indian must tell his neighbor, whose duty it is to take the message to his next neighbor; this goes on until all know about it. For this clay is regarded as public property, and every inhabitant of Gay Head who is willing to dig and help load the ship receives a part of the profit, which for these people is not small. A ton (2000 pounds) of this clay is sold for three dollars, and a man can, without much exertion, produce a ton a day. The ship which is now here loads approximately 90 tons.

Because I have searched so thoroughly the small area of my investigations that anything of interest which perhaps might still be here could be obtained only with much trouble and at great cost, I have this evening terminated my local collecting. I shall now somewhat put in order the collected objects, send a part of them from New York to Germany, and keep back another part for the start of the main collection. I will now visit again my old homeland, the Far West, and later on turn toward the South. I leave this place very satisfied and hope that those places which I plan to visit next will produce comparatively the same gain for me as Gay Head.

Wednesday, the 7th of August. After I had spent the morning until after ten o'clock settling my collection, I went on my way to the slopes where I had spent so many hours, in order, if possible, to make a small sketch of them, but before I arrived there I had a quite unexpected small adventure. During the night heavy dew had formed, and, so as not to get my feet too wet (my shoes had not dried out very much in the previous days), I took a longer but drier way where the grass was not so high, but I had to pass a narrow spot which was somewhat wet and had very long grass. I was almost through when I suddenly noticed that the middle of my foot hovered over a hideous snake and almost touched it; it is still a mystery to me that I did not step on it, but, sensing the danger at the right moment, I drew back as if somebody had unexpectedly poured a bucket of cold water over my head. Had I stepped on the snake I would have been lost because it would have wound itself around my foot and bitten me. The snake, far from being as frightened as I was, had coiled itself up ready to fight, and looked at me with glittering eyes. Quite surprised by the boldness of my challenger, I looked around for a stick but, alas, in vain because, as I already mentioned, there are no trees growing on Gay Head, and sticks and twigs are objects which have to be imported. What should I do now? I could not leave without having shown the snake that I had proved already to many of its peers that it would be best to run off when confronted by a hereditary enemy superior to them. Then my eyes hit upon a sort of butcher knife which I always carried with me, along with a bucket, when I went to the canyons. (The first I used to dig the objects which I found carefully from their resting place, the latter to scoop out water from the ocean for cleaning them.) So I took the knife and threw it at the snake but did not hit it because I could not dare to get too close to it. The knife fell not quite a foot away from the animal, which seemed to be very offended by it but in doubt as to whether it should revenge itself for the insult on me or the knife. I did not give the snake time to come to a decision, however, but hurled the bucket, which touched it only lightly. This made it so angry that it struck the bucket twice with full fury and bit at it. This attack on the bucket gave me the opportunity to get back my knife, which I threw a second time with somewhat better results, for it hit the snake on the back so that in first fright it drew back a little piece. I snatched the knife and the bucket back, but the snake again slid away a short distance so that I tired of the whole affair and went, though with great caution, on my way. When I arrived home at noon I was told that these snakes were very poisonous and seldom got out of anybody's way. No other name is known for them here, except that of the "spotted" snakes.

After I had climbed a rock partly submerged in the sea and had drawn my little sketch, I took a way which I had always bypassed earlier because of its steepness. With great effort I had climbed halfway up the canyon when I saw a small side canyon I had not noticed earlier, in which I soon saw the well-known conglomerate that had yielded so many beautiful objects. Although it was already the hour when my good old Indian hostess waited with lunch for me, I could not help but investigate for a while with the intention of repeating my visit toward evening and spending a few hours in which I could examine this little spot sufficiently. My expectations were not disappointed, and with tense hopes I again entered the place toward evening; for a long time nothing special showed itself until I finally saw, fairly high above me, a dark object. How pleased and surprised I was when shortly a big piece of the conglomerate lay before me, enclosing an organic object which at the beginning I thought to be a connected row of vertebrae! On closer examination it proved to be a piece of a colossal species of cane which seemed to me to come closest to the bamboo cane, and which had approximately a 7- to 8-inch diameter. It is completely petrified and shows very clearly the exterior and interior construction; the outer shell is divided in 2½- to 3-inch-wide rings and is comparatively thin. The interior part is interwoven with cells ¾ of an inch high and ¼ to ½ inch wide, of which many, despite their delicacy, are still beautifully preserved. After I had secured the find as well as possible, I found a second piece of the same species which, however, for the greatest part had been consumed by fire but which has a special value for me because, in the still wholly preserved cells, crystals have formed which, although small, are of great beauty.

✤ Holmes Hole, Boston

Holmes Hole, Saturday the 10th of August. From early yesterday until late last night, and again this morning, I was busy putting in order and packing my collection. Then, with the help of a half-breed, I put the things on a wagon and said farewell to Gay Head and my old hosts. The road was very bad and the weather very hot, therefore our journey went very slowly, but we also had to walk beside the wagon the largest part of the way; at least this had the advantage that we could refresh ourselves with the black raspberries which grew along the way. When we were still approximately 5 English miles from Holmes Hole, the man who carries the mail and with whom, as I remarked earlier, I had ridden most of the way to Gay Head, overtook us, and this man kindly invited me to take a seat in his small carriage, which I accepted with thanks. Safely arrived in Holmes Hole, I found my suitcases and everything I had left behind well taken care of. Many inhabitants of Holmes Hole came to see the objects which I had brought back from my excursion, but their curiosity could not be satisfied today because I had to postpone the unpacking until I had enough time to pack everything carefully anew, something I could not do in Gay Head.

Sunday, the 11th of August. This morning I went to the local Methodist church, a very friendly small building which on the outside is painted with white oil paint and has, like almost all houses in Holmes Hole, green window shutters. The floor is covered with beautiful rugs and the inside is, on the whole, very nicely furnished but without any ostentation. The sermon was very good, as was the singing of the congregation. The bigger part of the afternoon I devoted to my studies. Besides the aforementioned church this place has a pretty Baptist church and a Presbyterian church. Almost all local inhabitants as well as the ones on the whole island are more or less engaged in, or are participants in, whaling, whereby they seem to enjoy considerable prosperity.

Monday, the 12th of August. From morning until late evening I have been busy with unpacking and repacking my Gay Head collection. It contains in all 585 parts, namely 52 vertebrae of various animals; 19 bones of other parts of the body of various animals; 62 teeth of sharks and saurians of various species and sizes; 1 tooth and 1 incisor of a large, still-unknown animal not unlike the *Iguanodon*, which however was a carnivorous one; 3 incisors of a big saurian; 3 pieces of the exterior covering of saurians; 8 pieces of a very special conglomerate found in green sand with many bones and scales of fish; a large breastplate of a species of crabs or lobster unknown to me; 325 large and small pieces of crabs; 38 petrified shells from 3 species; 20 pieces of still-unknown petrifactions; 40 pieces of large and small parts of a colossal cane species which has not yet been described, as far as I know; and 12 pieces of the same cane, in which are, in which are found crystals. All morning long I received many visits from inquisitive persons, which, as one can indeed imagine, were anything but welcome.

Wednesday, the 14th of August. Because I was kept by adverse winds from departing on a small sailing ship yesterday, I had to make the journey today on a steamboat. At twelve o'clock I went with my things from Holmes Hole onto the certainly not large but quite new and elegant steamboat, which carried me in the time of scarcely two hours to New Bedford. Because of the ebb tide, the water was very low, so I had the opportunity of observing the dangerous spot which all the ships that make this trip have to pass. Approximately at the halfway point Elizabeth Island almost joins the mainland, and a dam of cliffs here connects the island with the mainland. Here is left free for the ships only a very narrow passage, not wider than 150 feet, through which the water presses with such force as one finds only in rapid rivers. In New Bedford my troubles started with transporting my two boxes, because the carriages which stand at the landing to pick up the travelers for the various hotels are nice big carriages, strongly built to carry many suit-cases, but not prepared to take on heavy boxes like mine. On the train to Boston it was the same, and in Boston, where I arrived at half-past five it was not much better.

The first thing I did, after having changed clothes and eaten dinner, was to look up my old traveling companions, Mr. Ludewig and his wife. I must also mention that during dinner, which in America, just as breakfast, is taken *table d'hote*, I had my seat beside Mr. Melancour, a Frenchman whose acquaintance I made on my first visit here. After a walk of an English mile I finally found my traveling companions, who were very delighted at seeing me again. After we had talked a while, Mr. Ludewig told me that another native Saxon, a certain Dr. Richter from Chemnitz, was staying at the same house, but I was not a little surprised when I found him to be my old acquaintance Counsellor Richter from Saxony, who was at the time of our first meeting a piano teacher in Detroit, Michigan, although he was very hard of hearing. After a pleasantly spent evening my friends took me home.

Thursday, the 15th of August. This morning I delivered a letter of introduction from Professor Silliman to Dr. Wyman.[1] To my not small surprise I met in Dr. Wyman, a man whose acquaintance I had already made earlier in London. We were therefore both delighted to see each other again and went to the museum of the local society of natural science, which has a very good, scientifically arranged collection. Unfor-tunately the hall in which it is exhibited is not big enough.

Mr. Melancour told me that a certain Mr. Prattford in Cambridge, for whom I had brought a letter from Privy Councillor Reichenbach of Dresden which I was prevented from delivering myself, wanted to make my acquaintance. Because Mr. Prattford is one of the most influ-ential people in Boston, I had arranged with Mr. Melancour to make a short visit in Cambridge, in which Mr. Ludewig and Dr. Richter participated. Every twenty minutes buses go from Boston to Cambridge, and in one of these we made our small excursion. The larger part of the way is on a bridge from which one has a very pleasing view. We were very kindly received by Mr. Prattford, and, after we had talked for a while, he led us into the botanical garden, to which no one is admitted without special permission. This garden is laid out with much taste and seems to be more a pleasure garden than a botanical. Everything was in full bloom, a fact that impressed me the more agreeably because when I left Germany the gardens were in their spring flowering, and this was the first garden of any importance I had seen in its summer dress. When we returned from the garden Mr. Prattford took us to the University library, which is housed in a very handsome, churchlike building, standing almost in the midst of the city on a grassy square which is adorned by beautiful trees. All around the building were inscribed in gold letters the names of those who had made significant contributions to the library.

✤ New Haven, New York

NEW HAVEN, Friday, the 16th of August 1844, late evening. This morning at half-past six I left Boston for New York. I had planned to go via New Haven, to show Professor Silliman my collection and hear his opinion about some of the things. Toward one o'clock in the afternoon we arrived in Springfield, where I again had a hard time with the boxes and my big heavy suitcase, because I had to transfer to a steamboat which was moored half an English mile from the railroad station. Everywhere I was told: "We are not equipped to handle freight!" There were very many passengers on the steamboat, but because we went down the Connecticut it was very fast and in 2½ hours we were already in Hartford, the destination of the steamboat. Very interesting but not without danger was the ride through the so-called Connecticut Rapids which are almost three English miles long and which one circumvents by using a canal with locks, when going upstream. But at the descent the owners of that ship would rather risk wrecking it and losing their passengers than pay the toll which they are charged in the canal. I am convinced that this trip would not be allowed in Europe. The river throws itself, like a mill-dam, roaring over rocky bottom; the little navigable water is in many spots so narrowly channeled that not a foot's width is left on either side, and it is really admirable the way those daring men maneuver the boat through. What makes this trip possible is that the water is crystal clear, and every threatening rock can be seen.

Before we came to this spot all of us had to move to either side of the boat, and all luggage, even the smallest, was put at the sides so that the two pilots necessary here could have an unobstructed view of each other. One of them stood in the back at the rudder, and the other was up front so that they could communicate by signs if danger threatened. The man in front had a kind of rudder of his own so as to help out at times. Everything went well, and I arrived safely in New Haven.

Saturday, the 17th of August. In New Haven it was my first business after breakfast to visit Professor Silliman, who lived more than one English mile from my hotel. He received me like an old friend, and after I had told him briefly how I had found everything and what I had brought with me from my excursion, he had his coachman hitch up to take me to my pavilion. He had also ordered another wagon which was to take my two chests to the geology lecture room, which he put at my disposition for unpacking the collection. During this job, with which the coachman had to help me, I found to my great joy that the objects had suffered no harm despite the throwing around of the chests. The professor was very much delighted with the collection, and in the afternoon he brought with him a friend in whom I met a very scholarly man, named Professor Shepard[1] of Charleston in South Carolina. He urged me to visit him on my way back from Alabama because he wanted to show me some places which would be very profitable for me. We made very interesting examinations and drew comparisons of my collection with objects in the local university museum. Among other things, we found that the three teeth I had uncovered in green sand, and which I thought to be saurian incisors, belonged to a colossal crocodile appearing for the first time in the paleontology of America. Judging by the teeth and the vertebrae I found with them it seems to be the largest animal of its kind yet discovered. After our examination was finished, Professor Silliman said to me: "Dr. Koch, you have thrown more light on Gay Head than all the others who have been there before." These words from this man were very gratifying because he himself, like some other great geologists, had furnished long reports about this highly interesting and noteworthy geological region.

Sunday, the 18th of August. Today all is quiet, even the steam carriages and the steamboats rest, and anybody who should by chance take it into his head to travel can be punished by law. This day of rest did not inconvenience me because I had a lot to do, especially with regard to cataloging my collection; therefore I have written almost all day, with the exception of two hours which I spent in church.

Monday, the 19th of August. All morning I was busy with packing my collection and with the transport of the chests to the steamboat freight agency. In the afternoon I went to bid Professor Silliman farewell. He had the kindness to take me to meet a lady who gave me much information about the region in Alabama where the big *Zygodon* is found, and who also presented me with two letters of introduction of great importance at that place.

New York, Tuesday, the 20th of August. The steamboat which left New Haven last night at ten o'clock brought me this morning at half-past three to New York, but, because at that time everything was still dark and the unloading of the freight did not start until daybreak, I could not get possession of my chests, which were buried deep below the others, until about seven o'clock, when I was on my way with them and my other luggage. The latter I took to my old quarters, and the former to a shipping agent.

This afternoon I visited for a short time the American Museum, where I was directed to a man who had a great quantity of bones excavated in Missouri. Unfortunately I did not find him at home, so I could not learn in what part of Missouri they had been found. There was, however, almost nothing of scientific interest among them, and everything was so damaged that one could not find much pleasure in the collection. They were mostly pieces from the antediluvian elephant or mammoth, and only a few of the *Mastodon giganteum*.

✤ From Troy in Pennsylvania to Cincinnati in Ohio

TROY,[1] WEDNESDAY, August the 21st. I departed from New York yesterday evening on the big steamboat *Albany*, which ran aground last night on a sandbar, as did the boat *Rochester* which left at the same time. Hereby we were an hour late when we arrived in Troy at half-past eight instead of half-past seven. Because of this one hour I lost twelve hours, for instead of being able to leave here by train for Schenectady this morning at half-past seven, I had to wait here until half-past seven in the evening, which cost me one dollar in unnecessary expenses.

Erie Canal, Thursday, August the 22nd. Last evening toward eight o'clock I went by train from Troy to Schenectady, which we reached after an hour and where our luggage was taken to a canal packet boat which was ready to depart and had waited only for us. I thought I would find only a few passengers here because many prefer to travel by train to Buffalo, which takes only 40 hours, while the packet boats need 75 to 76 hours. I was very much mistaken, however, because when we arrived the boat was so full of passengers that there was hardly a seat left for us. The reason might be the cheapness of the canal trip, for, while the journey on the train with subsistence costs about $13.50, the one on the canal is only $7.75. Right after our arrival the boat departed, and it took some time until all of us had paid their fare, at which point our names were written down; yes, it was almost eleven o'clock before the hanging beds of the ladies were prepared. While this was being done, all the gentlemen had to go up on deck because the ladies had taken over our cabin for the time being. Now our beds were hung, in three tiers and in three rows, on each side and in the middle. After that, each was singly called in and assigned his bed, all of which took until half-past eleven. At daybreak several began to talk, so that we did not get much sleep. Toward five o'clock we reached Utica where we changed to another boat, a change from which we profited because this boat was not only much bigger but also more elegant, and we were served better meals.

Friday, August the 23rd. This morning at eight o'clock we reached

Syracuse where we had to change boats anew and got a boat a bit smaller, but very many of our company left us also. It is worth noting that during my stay in New York one did not know what to do because of the heat, but during the night when I was on the Hudson it got so cold that the next morning all the travelers were seen in large topcoats. The cold lasted until this morning when it began to get comfortable again. All day yesterday we had a fire in the stove of the cabin, which we found very pleasant.

Saturday, August the 24th. This morning we came to Rochester where we changed boats for the last time. I had an annoying thing happen to me this morning; when I got up I bumped my head against a lamp so that oil was thrown all around me and on a letter to my wife and parents, which I was going to send the following day from Buffalo to Germany.

Buffalo, Sunday, August the 25th. We arrived at six o'clock in the morning and had to wait until seven o'clock in the evening because today, a Sunday, no steamboat left for the West. Buffalo has grown and improved a lot since I last saw it. Shortly after my arrival, I heard that again an accident had happened at the Niagara Falls. A young lady, who had come here to admire this great curiosity of nature, had on the so-called Table Rock bent over the declivity to pick a little tree branch, when some of the ground under her feet loosened and the unfortunate one fell into the 120-foot deep abyss. Her crushed corpse was found after some time and taken to Buffalo, where she now lies in a coffin in the U.S. Hotel. This is the second victim this year who found destruction on this rock; in the spring a young man met with a similar accident.

As I just see from St. Louis newspapers, it seems to be still the fashion in Missouri that a certain class of people take possession of the law when in their opinion the court is too slow or the guilty ones are not sentenced. This is shown by the following occurrence. In a small town in Missouri a man was sentenced to hang as a murderer. The execution was about to happen, and a great number of curious onlookers were assembled to watch the tragedy. The delinquent stood under the gallows when (the reason is unknown to me) the Governor found it proper to postpone the execution for two months. This so angered many of the spectators that they joined to hang the man at once, themselves. This, however, was thwarted by a company of soldiers and the better minded citizens, and the criminal was again returned in chains to prison. Approximately only two weeks had passed when the above-mentioned party, armed, went to the jail, broke open the doors, put a rope around the neck of the delinquent, who was lying in chains, and lifted him up. To all appearances he was dead before it came to the actual hanging, but he nevertheless was dragged to a nearby tree and was hung there for a while.[2]

Erie, Monday, August the 26th. This morning at four o'clock I arrived in Erie after a nine-hour steamboat ride from Buffalo; we had head winds, otherwise we would have made the trip in seven hours. To my delight two carriages waited at the landing place, which was a bit remote from the city, and I rode in one of them to the big Hotel Reed House, where I caught up on the sleep missed on the steamboat. After nine o'clock I visited my local relatives and learned that my good old uncle,[3] who had been a preacher here for many years, had died in May. How great the sympathy was which was displayed here at his death is shown by the fact that more than 1,000 persons accompanied his remains to the grave. I found Erie very much improved in the 14 years of my absence and visited some old acquaintances. The weather had been so cold for several days that there were fires in the rooms almost everywhere.

Tuesday, August the 27th. Because it rained all morning, I was able to work in all quiet on a geological report about my last discoveries in Gay Head, which I had promised to Herr Professor Silliman for his journal. Toward evening I visited my relatives, by whom I was received very warmly, and was invited to live with them during my stay in Erie. I accepted this invitation because I did not know whether I would ever see them again. That same evening I had my luggage taken to them.

Friday, August the 30th. Of the events of the last days the one most worth mentioning is that I discovered here some very interesting petrifactions in a schist layer, which is rare; among them were four of remarkable beauty. Yesterday I had a chance to see the canal which connects, from Erie on, Lake Erie with the Ohio.[4] It was started 16 years ago, and it will only next fall be completed to such an extent that it can be called finished. But this fall a boat has to pass through, or the company which is building it will lose its rights.

Cleveland, Saturday, August the 31st. Yesterday evening, at seven o'clock, I traveled on the steamboat *Buffalo* from Erie to Cleveland, where I arrived this morning at half-past five. It was a beautiful, calm night and therefore a pleasant trip. The steamboats and other ships are now forced, when they come to Erie, to make a long detour. Very soon, however, this will no longer be necessary because they are hard at work on a breakthrough by which a straight entry into the Erie harbor will be aimed at, whereby the city will gain as much as from the already-mentioned canal.

My first business in Cleveland was to find a canal boat on which I could travel on the Ohio Canal to Portsmouth, a city which is a distance of 307 English miles from here and which lies on the spot where the canal connects with the Ohio. Such a boat was recommended to me on the steamer, but I found it so dirty and unattractive that I left immediately and went to a rather decent looking second one, which, as I was told, would not leave until four o'clock in the afternoon. I was bored until that hour all day because I could not get to my things to write or read. But even now no preparations were made for our departure; why, the captain even told me it was dubious whether he could depart today, and he offered to take me to another boat. As disagreeable as this delay was for me I had to submit myself to it. The boat to which I was transplanted was much more comfortable than the one I left, and it was supposed to leave at eight o'clock in the evening. We actually moved at that time a few hundred feet, but came to a stop again, and the word now was that we would leave during the night.

Sunday, September the 1st. At nine o'clock in the morning we were still lying at the same spot; the horses had been harnessed for two hours, and all of us soon lost patience. The reason for the delay was no other than that our captain was awaiting three more passengers, after three passengers had already come to us during the night off a steamboat from Buffalo. In a certain respect I welcomed the addition, because the mosquitoes tormented us miserably and we were almost too few people for the great number of hungry and bloodthirsty animals which could now, with the increase in our company, distribute themselves a bit more.

Ohio-Canal, Monday, September the 2nd. Yesterday at half-past nine our journey finally went on. Among our passengers was one who had, nine weeks earlier, broken his left leg on a canal boat. The bridges which lead across this canal are very low, and one has to be very careful when one is on deck. When an unusually low bridge comes up, it is very easily possible not to notice it in time when the passengers are involved in an interesting discussion. It was such a bridge which broke the leg of the poor man. In greatest pain, and plagued by wound fever, he had to be brought back, and now he was for the second time on this trip, and, as can be indeed expected, he was one of the first ones who ran to safety as soon as the warning call "A bridge!" sounded. The change of climate was already very noticeable, because it was so hot again that one had to get his summer clothes out and find a shady spot.

I was satisfied in every respect with the boat on which I traveled; not only was our cabin very elegant and the meals very good, but the captain as well as his subordinates knew their business. I paid only $6.50 and no tips for the entire journey, which took six days. Toward evening we stopped in Massillon for ¾ of an hour to unload goods. Massillon is a nice little town, which consists for the most part of fine, big houses and wide streets and gives evidence of, in relation to its size, much affluence and a brisk business life.

Tuesday, September the 3rd. The heat had increased more and more so that during the hours when the sun stood high all withdrew to the cabin. All the more beautiful, in contrast, were the mornings and evenings, the first of which we had full opportunity to enjoy because our breakfast was ready early at six o'clock. We passed several coal mines today which are worked in a very simple way. At the foot of a rocky hill an entrance is made four to five feet wide and high enough so that a man standing erect can pass through without bumping his head. This opening goes rather horizontally into the mountain, where it meets the coal deposit, and from here the bituminous coal is taken by ordinary wheelbarrows to the canal, only a few steps away, and thence is transported by the canal boats.

Wednesday, September the 4th. This morning we passed a romantic region; the cliffs whose base the canal passed close by had much similarity to the ones of our beautiful Saxon Switzerland,[5] except that these were covered with red cedars and deciduous trees and were only approximately 100 feet high. Toward evening we came over an artificial lake which is more than two English miles wide and seven English miles long. Through the middle is built a stone dam as a path for the horses which draw the boat. The purpose of this lake is to supply the canal on both sides with the necessary water from a river called the Licking,[6] which is situated at the greatest height that the canal reaches. It looks as if this lake abounds in fish, because everywhere one saw fish jump out of the water. There was also, however, a large mass of mosquitoes, which made me very anxious about the coming night, but we had taken all precautions to keep them away from our cabin.

Thursday, September the 5th. Toward evening we came through the town of Circleville.[7] For a long time I had cherished the wish to see this place and its environs because here is to be found irrefutable proof that Noah's flood, or the deluge, as it is called in the Holy Scriptures, also flooded America—something which has often been questioned by critics and which is still considered doubtful. Yes, even more one finds here near Circleville one of the best proofs that America was inhabited by human beings at that time of terror, because from this diluvium were unearthed human remains which have been preserved until our time, to appear as witnesses in our criticizing age. Although I had mentioned this fact in my evidence for the habitation of America before and during the time of the flooding of our earth, I was very gratified to have found the opportunity to see this region myself and to make the personal acquaintance of Mr. Atwater,[8] to whose history of Ohio I have referred.

In Circleville itself and its environs one finds elongated hills consisting of gravel, which show themselves to be from the diluvium by the sea shells they contain. Among the gravel, which is much used for filling in roads and so forth, were discovered in 1824 seven human skeletons whose skulls were, with the exception of three which were badly damaged, very well preserved. Unfortunately, however, they were pilfered from their discoverer, Mr. Atwater, a few years ago.

Friday, September the 6th. During the day we passed several sawmills in which no wood was cut, but a beautiful, fine sandstone, which is also quarried here, is cut into slabs 2½ to 3 inches thick. The saws are fixed so that eight slabs can be finished at one time.

On the Ohio, Saturday, September the 7th. Last evening we came to the Ohio, but we could not cross the river because the night was too dark. The horses, which had pulled us until now, were unharnessed. We had four horses of which two alternately pulled while the other two were carried; on all the local canal boats there is a stall for the resting horses. Every six hours, day and night, the horses are changed this way, and approximately 55 to 60 English miles are covered daily. We would have liked to stay on the boat until we could board a steamboat, and therefore offered to pay for breakfast, but the sailors and the innkeepers here are in collusion so that we absolutely had to go to an inn. We had hardly taken our breakfast when a steamboat arrived from Cincinnati on which we took quarters at once and with which we arrived at two o'clock in the afternoon in the town of Maysville, which lies close on the bank of the Ohio in the state of Kentucky. We learned that the boat would spend some time here because a lot of freight had to be unloaded, and we used this time to climb the high rocky banks of the river although the heat was very intense. Because of some scattered stones my attention was drawn to the fact that the local region must be rich in petrifications, and I found my expectations almost surpassed because the mountains were full of beautiful shells, among which I also discovered some coral species. In a hurry I collected as many of them as I could take, and, because I had to carry my treasures in my hands, I greatly astonished the local inhabitants.

Cincinnati, Sunday, September the 8th. This morning at half-past six we arrived here and had to take another steamboat scheduled to leave in two hours, but which unfortunately did not depart until four o'clock in the afternoon. I tried to see several old acquaintances but did not find them at home and so spent most of the day partly in unsuccessful visits, partly in waiting and tarrying. However I still had the opportunity of seeing the tomb of the late president of the United States, General Harrison, who lies buried a few hundred steps from his house in North-Bend on a hill at the Ohio. It consists of a simple obelisk surrounded by a railing.

✤ Louisville in Kentucky, Jeffersonville in Indiana

LOUISVILLE, MONDAY, the 9th of September. This morning at six o'clock I arrived here, the next goal of my journey. The first thing I did was to go to an old fossil collector whom I had known previously and who gave me advice as best he could, but to be sure I walked to New Albany, which is four English miles from Louisville, to visit Dr. Clapp,[1] the best geologist in this region and the owner of an important collection of local fossils. I had a few lines from Professor Silliman to him and was received cordially. The result of my inquiry was that to be closest to the place where I intended to collect, namely the falls of the Ohio, I would have to go to Jeffersonville.

Tuesday, the 10th of September. My expectations concerning the local fossils I found quite satisfied, if not surpassed. The shells and corals are frequently so beautiful and perfect, as if the animals whom they once served as houses and covering still lived in them. Also their variety is considerable. Although the place of discovery was only one English mile away, and I left as little stone as possible on the collected fossils and took only the best, I had to carry so heavy a load that my arms were quite tired when I arrived home. As I could not take all my selections away at one time, I had to make two trips.

Jeffersonville, Wednesday, the 11th of September. Because yesterday I collected fossils which appeared in the upper stratum of the local limestone, I now went to work on a lower stratum which is considered part of a still older secondary formation and which contained almost more petrifactions than the first. Indeed, I made the observation that the whole bed of the Ohio, as far as the falls of it stretch, that is four English miles, is an almost uninterrupted coral deposit. The corals existing here are of 45 to 60 different species, and many of them are of an extraordinary size. In some layers is also found a conglomerate which consists of iron with shells and coral. A rather special phenomenon is that many of the shells, as well as the corals, lie loose on the rocks in which they once were embedded but from which they have been apparently torn out by the water.

Thursday, the 12th of September. Since my arrival there has been an almost unbearable heat which seemed to increase daily, and the places where I worked offered not the slightest shade. I had to be very careful, therefore, to preserve my health, which, thank God, has not yet been troubled, although many of the local inhabitants were laid low by fever, and others wandered around like spirits risen from the grave. Also mosquitoes were here in such large numbers that in the evening one could neither write nor read. To my consolation the beds were surrounded with nets so that at least at night one was protected from them and could sleep in peace.

Friday, the 13th of September. My collection of local fossils has grown now to a considerable size, and I think I can say that only a few species exist here of which I do not have at least one specimen. The weather is still very hot, and every day the Ohio drops considerably causing the steamboat service, of which I want to make use again, to be impeded.

Saturday, the 14th of September. This morning some inmates of the local penitentiary of the State of Indiana [2] worked not a hundred steps from me; they were employed to break loose flat platelike stones which are found here in great mass and are easily obtained. It was horrible and sad to see those people. On their deathly pale, ashen faces sin had pressed its mark. Their head covering is the worst I have ever seen and is not enough to cover the close-shaven head; their shirts contrast in color only a little or not at all with their gray suits, of which one half is dark and the other light, and around their feet are wound heavy chains. From a distance of 10 to 15 steps they were watched by a guard who, with a cocked rifle and a grim face, sat on a rock.

I had just stood still for a moment looking at the group and absorbed in contemplation when an Irishman joined me who, with some companions in another spot not far from here, had dynamited rocks to burn lime which is shipped in great quantities from here. After a simple greeting he asked me what I was going to do with the stones which I was sorting out so laboriously, since they were too small to be used for building houses. I answered him that they were collected for scientific purposes and then called his attention to the pitiable appearance of our neighbors. Ah, he replied without a moment's reflection, you must not forget that these people did not get into this situation building churches or furthering religion.

Sunday, the 15th of September. All America is now in the highest degree of excitement because of the impending election of a new president.[3] Not only in all places where two or more people get together is there talk about this subject, but also even late in the evening one hears yelling on the street: "Hurrah for Clay!" or in opposition, "Hurrah for Polk!" Yes, one cannot walk in the streets without being called to in that manner by children, even those who can hardly speak clearly. Every day small or large political meetings of both parties take place, and in the newspapers one finds almost nothing but political reports. On the whole I am not very much interested in this affair, and yesterday evening was the first time that I attended one of these political meetings, which was arranged by the local followers of Henry Clay and took place in the market hall. The meeting, in which ladies also participated, was opened by the singing of some songs which praised Henry Clay and the states especially loyal to him and ridiculed the opposing party. Then followed a speech, lasting 1½ hours, by a lawyer named Wolf from Louisville, who very well explained the consequences which would follow the election of one candidate or the other for president, and then called to the crowd's attention the difficulties into which America had been put by electing the wrong people and from which it had just barely recovered. After the speech a few more songs were sung, and the meeting then broke up.

This morning I visited the local Methodist church but I was only slightly, if at all, edified by the visit. After lunch I went with a London gentleman whose acquaintance I made here to the local sulphur spring, a public garden one English mile from here, which is laid out in English style and is much frequented by the inhabitants of Louisville.[4] In the evening I went to the local Baptist church, actually more to escape the mosquitoes than anything else, but to my joy I heard here a very nice sermon and went home very uplifted.

Monday, the 16th of September. Because I was very anxious for Dr. Clapp of New Albany to see my collection before I packed it, I went to him after breakfast, on my way to New Albany, which lies at the lower end of the Ohio falls just as Jeffersonville does at the upper. Both places are connected by steam ferries with Louisville, which lies on the other side of the river somewhat below Jeffersonville. Because the larger steamboats cannot pass the falls except at unusually high-water level, they must pass from Louisville through a canal, three English miles long, for which they have to pay a monstrous toll, amounting to $100 to $200 per boat, according to size, and which has to be paid whether the boat has a full load or not.

After I had gone two-thirds of my way I met Dr. Clapp, who told me, after a friendly greeting, that he was about to make a little botanical and geological excursion to the Ohio falls. He is not only the best geologist for miles around but also the best botanist, and, because he occupies himself exclusively with it, his authority in respect to the classification of natural objects is very valuable. We agreed that the doctor should visit me in the afternoon, and I went home to put my collection in order so that it could be easily examined.

Dr. Clapp was not a little surprised to find my collection so rich in beautiful and rare objects; indeed, he designated some among them as not yet scientifically studied, and some were even new to him who had collected here for 20 years. Many of the objects the doctor himself had recently described. He was so interested in the collection that he promised to come again.

Tuesday, the 17th of September. The falls of the Ohio are divided almost in their entire length into two parts by a rocky island, two English miles long, of which not much can be seen at high- or regular-water level, and which is again split by many gorges into smaller islands. For several days I had the wish to inspect those islands and had waited only for a very low-water level. This morning I had two boys take me there in a boat which they handled very skillfully, which was absolutely necessary because in the middle the stream rushes down with such force that it creates foamy waves several feet high which can quickly turn over or sink a boat that is not handled properly. I found some very nice and rare petrifactions here, and, because I do not care to repeat the trip on the Ohio falls, I worked so strenuously that I developed blisters on my hands and was bathed in perspiration from the great heat. Very remarkable is the fact that the islands I have described are covered with ½-foot to 1-foot-wide cracks, like ice in hard winters but on a larger scale.

Wednesday, the 18th of September. Because I spent the whole day sorting and packing my collection, I used the warm moonlit evening to take a swim in the Ohio. The Ohio falls afforded a magnificent view in the moonlight, and the hundreds of lights in Louisville, lying opposite, added much to heighten my enjoyment. The stillness of this beautiful summer night was only interrupted by the noise of the falls sounding like faraway thunder.

Friday, the 20th of September. This day too was almost entirely given over to the sorting and packing of the objects in my collection. Yesterday and this evening many cannons were fired in Louisville. Last evening it was in honor of a Democratic speaker who spoke to his party, but the significance of tonight's festivities I could not find out.

Saturday, the 21st of September. Yesterday Dr. Clapp spent several hours with me to help classify the petrifactions in my collections. This morning I visited him again to look at his collection and to find out which objects I lacked. On my way back after lunch I did find several petrifactions which until now I had missed in my collection.

Sunday, the 22nd of September. Today I visited the local Baptist church, whose preacher pleased me very much. It is very small, like all churches in this place, and has the peculiarity that directly behind the pulpit hang two rather large maps, one showing the two hemispheres and the other, Asia. Today was the first time I ever saw maps in a church, and I thought they had perhaps been put there for the use of the Sunday school. But I observed later that the preacher made very good use of them when, several times during the sermon, he pointed out on the maps places which appeared in the text he had chosen.

✤ Charlestown and Jeffersonville in Indiana

CHARLESTOWN, MONDAY, the 23 of September. My host in Jeffersonville had, ever since my first arrival in his house, told me much about the great number of petrifactions to be found in Charlestown. But, because I did not wish to exchange the certain for the uncertain, I have stayed until now at the falls of the Ohio, where, from morning to evening, I have found enough to do. During my last visit with Dr. Clapp in New Albany, however, I saw some very nice petrifactions which are not found here; I believe they have only been found in Charlestown. This induced me at once to decide to go there, especially since the place is only 15 English miles from Jeffersonville, and three times a week a stagecoach travels between the two places.

So as not to lose a moment of my now so precious time, I got up this morning at daybreak and worked at sorting and packing my collection until the bell called me to eat, and I had hardly time to pack a few necessary things in my travel bag before the stagecoach stopped outside the door to pick me up. The fare is very low; for the 15 miles it is only half a dollar, but it is not very well set up for comfort. The stagecoach is a lightly built covered wagon with three seats, each for two persons, but not more than five people can travel because half of the front seat is occupied by the driver. But the team of horses consisted of two beautiful sorrels with which no exalted personage would have been ashamed to ride. Affixed to the coach was a special device with which the driver was able to brake the coach immediately and just as fast let loose again without having to get off his seat; this device is now also very common in Europe. This saved us much trouble and time, because as short as our way was, leading us almost continually through forests, we had to pass many hills or small mountains were braking was necessary. My travel party consisted of a young man and a woman who, except for a question to the driver, did not speak a single word during the entire trip.

Tuesday, the 24th of September. This morning I made my first excursion here. I went in a northerly direction from the city, and no sooner had I passed the last house than I came to a small brook which was almost completely dry and which traversed the road from west to east. I followed this brook upward and in only a few moments I discovered several very interesting petrifactions which, however, were not contained in the rocky banks of the the brook but in a stone mass, broken in little pieces, which was mixed with sand and was washed here from time to time by the force of the water. This caused me to suspect that the rocks from which the petrifactions came must not be far from here. My conjecture did not mislead me because, as I followed the course of the brook which wound through a hornbeam forest and then to a place where it took its course through a cornfield, here were the rocks

I was looking for. I found here among other splendid fossils a magnificent large *Euomphalus*, then a nice shell belonging to the Teredo family, as well as an almost complete *Pleurarynchus*, all still quite unknown. The *Pleurarynchus* is a creature which only recently has been counted among the mollusks. I am still in doubt as to whether this is right because, although the front of the animal looks like a mollusk, the opposite end terminates in a wide, flat, hornlike tail. This tail, to the best of my knowledge, is found on no mollusk species. Dr. Clapp told me that so far only one example has been found that equals mine in perfection.

In the afternoon I followed the same small brook in the opposite direction; I found rocks enough, but they contained either no fossils or only some which I did not want. I finally came to a place where lime was burned, and in searching around the quarry I was delighted to find a crystal formation of exceptional beauty which is of great interest because it is so similar to petrified wood that for a long time it was thought to be just that. I had in fact already found a few pieces of crystal at the Ohio falls, but none so beautiful as this one. Half an English mile farther on I came to a mineral spring spouting from a rock over which was built a small wooden shed to shield it from the rainwater. I drank a few swallows of this water, but it had such an unpleasant taste that despite my thirst I did not want any more of it. The water contains, besides other mineral substances, a large quantity of Glauber's salt and acts as a strong purgative. On the way back I went over some rocky hills with many petrifactions, but they were like the ones in Jeffersonville. Back in Charlestown I met my landlord, Mr. Green, from Jeffersonville, who very kindly offered to show me this evening a very interesting place, which invitation I accepted with thanks. This spot lay also on the banks of a little river flowing partly through town but now dried out. The right bank as well as the riverbed consisted of limestone with innumerable fossils. Especially remarkable among them was a small species of sea urchin, of which we found several examples this evening. Dr. Clapp had drawn my attention to it as a fossil known to have been found only here.

After supper, a meeting of the Whig party was held in the courthouse. A young German appeared as speaker, first in German and then in English. Although he had no gift of oratory, he called the attention of his listeners to many political truths, which received repeated applause.

Wednesday, the 25th of September. A tailor, who has been living in Charlestown for 30 years and who had promised to show me a place rich in fossils, led me today to this spot. It was a rocky riverbed, and like the others now without water, over a mile from the city. I was not very satisfied at the beginning of our excursion when no petrifactions showed, because only now did I become aware of the main object of the tailor, who wanted to find out where coal was to be found here. He made many excuses and promised soon to take me to another place which I certainly would find satisfactory. Accordingly I showed him where to dig if coal should be here at all; shortly afterward we did come to a place where I found very nice fossils, among others the head of a rare *Pentremites,* and so at noon I parted, very satisfied, from my old tailor. In the afternoon I went again to the spot which supplied me yesterday evening with so many beautiful objects and had promised more. I was not mistaken because by evening I had collected so many rare objects that my boldest expectations were exceeded. Among other things, I discovered ten of the largest mollusks of a rare species, of which one alone would have delighted me. In the evening there was again a political meeting in the same building as yesterday, but of the opposite party, and, strangely, the speaker again was a German, namely a Dr. Holland [1] from Louisville, in Kentucky.

Thursday, the 26th of September. My collection is so constituted that I have decided to add to it now only the rarest objects, and because the local place of discovery, where I spent a large part of today and again collected with excellent luck, cannot henceforth provide me with anything new, I shall go back to Jeffersonville tomorrow.

Jeffersonville, Friday, the 27th of September. Last night I had a small red spot on my left knee which hurt a little. This morning it somewhat hindered my walking, and as I arrived here toward twelve o'clock my knee was so swollen that I could get out of the stagecoach only with much difficulty, and then I became so lame that I could hardly limp from my room to the dining room. Yet I will not complain in any way about it but on the contrary thank God that nothing worse confines me to my room, because not only here, but also in Charlestown, many people are still laid up with fever, and several have died from it, while I, despite all great exertions, enjoy excellent internal health and am among the first and the last at the table. Besides I have a lot to do with arranging and packing and, moreover, a lot to write.

Saturday, the 28th of September. My knee has not improved, and I worked the whole day.

Sunday, the 29th of September. The most beautiful day followed a heavy rain, but because of my knee I still have to keep to my room.

Monday, the 30th of September. This day also spent working in my room. This evening there was again a political rally in the local market house, and the good man who delivered the speech got so excited that I could hear him at my desk. Again, too, I heard cannons fired off in Louisville, which might have proclaimed some political news. But the scene soon changed in Louisville, because the cannons were silent and anxious alarm bells sounded and announced a fire, but the speaker in the market house did not let himself be put off but continued to shout.

Tuesday, the 1st of October. The beautiful American fall weather has now set in; it is neither too warm nor too cold, and I who have been confined to the room for three days sometimes get up from my work and, like a prisoner, look out the window. Oh, how I yearn to enjoy the weather and, even more, to be able to walk to the falls of the Ohio, which is now lower than it has been for several years and would open such a field for me as is not often found. God willing, I shall be able to limp tomorrow evening to the Ohio falls, as my knee is getting better.

Wednesday, the 2nd of October. I was not disappointed in my expectations of yesterday. This morning it seemed indeed rather risky to make an excursion with my lame leg, but, resolving to turn back in case the trip was not feasible, I took up my hammer and chisel, and, using my umbrella as a cane, I went on my way. Very slowly, I limped first to an old man named Parkly, who lives directly at the falls of the Ohio and who, although he possesses no theoretical knowledge, has collected during his life many fossils here and has already done me some favors. The poor old man had been sick since I had seen him and still felt rather weak. I told him therefore that if he wanted to keep me company we would match each other because both of us would have to walk very slowly. The truth of my remark mainly induced him to keep me company, and we let a young man who was fishing on the bank take us over to the island which I had visited with luck once before. Already I had found and collected more than 60 species of the rarest primeval corals and shells, of the choicest examples of which some had been first discovered by me, and now I was walking on a continuous layer of prehistoric corals of the most varied sizes, which, frequently interlaced with one another, formed curious shapes and often appeared like beautiful large flowers surrounded with garlands of leaves. One who does not have occasion to enjoy this view can best get an idea of it if he imagines the ice flowers and leaves on frozen windowpanes on a large scale and spread out over several English miles. I found a coral stem which formed a circle with a diameter of ten feet, although the branches were not over half an inch thick. Some corals run helter-skelter like tree roots; many have the form of garlands or resemble honeycombs so much that the local inhabitants took them to be petrified wax, petrified honey, or petrified wasps' nests; others, again, have great similarity to horns and are known by the name of petrified buffalo horns; still others look like the blossoms of the aster or the so-called horsetail and are often so small that one can hardly see them with the naked eye. In the course of time, the water has washed off the limestone, and the much harder petrifactions consequently lie raised on the stone slabs, more splendid than the most beautiful sculptured adornment; the beauty of it is made complete by the fact that the stone is gray-blue, and the petrifications look black. Enough, it is a truly enchanting view which indeed I would like to have described by a more skilled pen. This is, as I want to express it, a subterranean magic garden which is open only for quite a short time every four to five years at the most, as it is now for the inquiring geologist, and where there are uncovered for science many new species of those wondrous creatures which might well belong to the first works of creation, for they come from a formation which is even below the coal.

Thursday, the 3rd of October. Again today I came back home from the Ohio falls heavily laden with many fine specimens, and among them again some quite new and unknown. I had hardly arrived home when my friend Dr. Clapp from New Albany visited me to help me classify and arrange the collected objects, which took almost the whole afternoon.

Friday, the 4th of October. Because I had to walk a distance of three English miles yesterday and had to carry home such a heavy load, I put my hammer and chisel in a hidden place, as I had done on similar occasions before. This morning I walked, to gain time, at a sharp pace to the place where I had been collecting, and I was not a little surprised when I did not find my instruments. At first I thought I had missed the right spot and looked around for a whole hour in vain. Now there was nothing left but to borrow, and, because I knew that one English mile from here an old man had tools like those I needed, I went to him without delay and found him very willing to help me in my predicament. This noon I went home a half hour earlier than usual, and by two o'clock in the afternoon the local blacksmith had made me some new instruments, with which I found so much this afternoon that I had to leave a part until the next day because I could not take all the objects away in one trip.

Saturday, the 5th of October. This day was again one of those in my life of which I can say with good reason that if our infinitely good Father in heaven had not taken me under his special merciful protection I either would not be among the living or surely would be a wretched cripple for the rest of my life.

I had this morning as always collected with much luck, and I went right after lunch, cheerful and bright, to the Ohio falls to ride over to the island which I had visited several times. At the crossing point I met my friend Dr. Clapp, who was on his way to bring me a letter of introduction. It did not take much to persuade the doctor to go to the island with me. We arrived there safely and found many interesting things, whereby Dr. Clapp called me fortunate because of my quick grasp of the situation and my sharp eyes, and we sat down, pleased, in the boat in which two young people had just arrived to pick us up. To expose us to less danger they went a little farther down the river than usual. We were about to land when a shot was fired and a quantity of buckshot passed so close before my eyes that it caused me to blink involuntarily, upon which the young man who sat not far from me became so frightened that he jumped up. Eight pellets were embedded in the sides of the boat alone, and how many went around us we could not know. The shot came from a careless boy who stood on the near bank and had shot at a sandpiper. When we called the boy to account for his thoughtlessness, he replied that he thought we could have waited until he had shot, since he had had to wait long enough, and the bird was just about to fly away. To my question as to whether he knew what would have happened to him if he had killed one of us, he answered calmly: "O yes!" The boy deserved severe punishment, but nobody wanted to have anything to do with him.

Sunday, the 6th of October. This evening I went to the Methodist church, which was very full because a new preacher was appearing here for the first time.

Monday, the 7th of October. Yesterday I obtained the seed of a very beautiful garnet-red flower which is very long-blooming and belongs to the hibiscus family. This evening I terminated my collecting here because I now have everything that can be found here, and I can say without boasting that no collection of local objects is equal to mine. Still this morning I found an example of a truly splendid petrifaction about which I am in doubt as to how to classify it; it comes closest to the *Cyathophyllum* family; it has quite the shape of the bindweed. Its upper diameter is more than 6 inches, and its height approximately the same; the specimen is so well preserved that under the magnifying glass one can see its beauty very clearly.

Tuesday, the 8th of October. Because I needed some trifles I could not get here, I went this morning by steam ferry to Louisville, where in some streets it was very lively because this week not far from the city horse racing is taking place,[2] this being a favorite amusement of a certain class of Americans. I spent only a short time in Louisville and then worked until after eleven o'clock classifying and packing my collection.

Wednesday, the 9th of October. All day long without interruption I was occupied with putting my collection in order. Toward evening, on both sides of the river, cannons were fired to proclaim happy news from the East for the local and Louisville Whigs.[3]

Thursday, the 10th of October. Today a big *Volksfest* was held here, known as a barbecue; it is held at the time of the election of a new president by both parties at different places and different times as they see fit. Its main purpose is to assemble a large mass of people at one place and to present speeches about political matters. For this, as already mentioned, a date is set on which, at a selected shady spot, a noonday meal is prepared in the open for thousands; all who wish can partake free. Usually after the meal skilled orators mount speakers' platforms which are far apart from one another, so that one does not bother the other. Those who are fond of dancing are also taken care of, for not far away, also in the open, is a large band which plays for them free. At another spot, songs are sung by large groups which ridicule the opposition party and, it goes without saying, praise their own to high heaven. The local barbecue was held by the Whig party half an English mile from the city in a beautiful public park which I have already mentioned earlier. Already early in the morning the whole city was on the move, and banners with the names Clay and Frelinghuysen waved not only from poles almost 100 feet high but also from the windows of houses; one banner was stretched across the main street from one house to another. To perpetrate a prank, some young people had during the night cut the lines which held the banners of their opposition, and all of a sudden there appeared a large flag of the opposition party over the main street. On the one side it showed the American eagle, on the other, one star for Texas, and in the middle the names Polk and Dallas.[4] From all streets came wagons with large and small flags and standards with various slogans. Three wagons appeared, each carrying on a pole a small house with a live raccoon, the symbol of the Whigs; on the first the animal slyly looked out the window, on the second it sat in the door, and on the third, one saw it on the roof. Toward eleven o'clock a company of volunteers with fifes playing and drums beating came marching in from New Albany. They wore quite a peculiar uniform, very much like the dress of the Indians, I think in allusion to the name of the state of Indiana. They were followed by a great many wagons with ladies and gentlemen from New Albany. Each wagon had some kind of flag or something similar. Then followed processions from other neighboring communities decorated in the same or a similar way, and at the end were the Jeffersonville Whigs. The parade stretched from the city to the barbecue place, and I believe that 5,000 to 6,000 people were assembled here to eat, where, as could be expected, there was no little commotion. In the evening two more speeches were held in the local market house, closing the festivities.

Sunday, the 14th of October.[5] Last Friday and Saturday I was so busy writing and packing that on both days I could not lay myself to rest until after eleven o'clock in the evening.

✤ Madison, Dupont, and Jeffersonville in Indiana

M ADISON, MONDAY, the 14th of October. All last night it rained, and for a long time I was undecided whether or not I should start, as I had intended, my journey to Madison, but time pressed and because I had, at present, nothing more to do in Jeffersonville, I decided to go. Yet before I departed, my old friend Mr. Bright returned from a trip and hastily wrote a letter of recommendation to one of his local sons who is lieutenant governor of the state of Indiana.[1]

I took the steam ferry to Louisville, whence I continued my trip on the mail steamboat *Pike* and arrived here in the afternoon. The banks of the Ohio were in their most beautiful fall decoration and looked glorious.

Wednesday, the 16th of October. Last night I was so exhausted that I was not able to write even a few lines in my diary, and I did it this evening with great effort so as not to cause too big a gap. I was so tired that not only my fingers and hands but even the soles of my feet hurt, but I had the great comfort that I had received the richest reward, since the objects I found here had already exceeded my boldest expectations, not only in their beauty but also in their rarity and novelty.

The same formation exists here as at the falls of the Ohio and in Charlestown, but it contains many fossils which do not exist there, as well as several crystals. A few years ago a railroad was laid out here which should have connected the local place with the capital of Indiana. In the course of this work a rocky mountain one English mile from Madison was cut through to a length of 150 feet. This resulted in an excellent geological section, and, although I only enjoyed the gleanings of the uncovered fossils, I obtained through diligence and effort several rarities which the earlier collectors had overlooked. Particularly it was the smaller, finer corals, which demand the greatest attention and which are usually quite neglected, because everyone reaches for the big items that catch the eye, and even many collectors confine themselves to objects which they can get for a trifle from a quarry man.

The weather yesterday was very unfriendly; from time to time a cold, drizzling rain was forced into a very strong wind which whipped the whole Ohio, on which Madison lies, into small waves; today it was a little better, but still far from beautiful. The city of Madison, which numbers 3,000 to 4,000 inhabitants, is very romantically situated; on three sides it is almost directly surrounded by rocky hills covered with forests, and on the fourth side it borders on the Ohio, whose opposite banks also consist of hills. The streets are very wide, to a large extent planted with trees, and the houses are built tastefully; some are even quite beautiful.

Thursday, the 17th of October. Despite the continuous rain I did not let myself be deterred from my research work. In the morning it went tolerably; I only got wet from above because I walked, climbed and stood, as was necessary, on broken pieces of rock. In the afternoon I almost became impatient for I worked at a spot which, to be sure, contained the most wonderful fossilized shells, but the rocks in which they were embedded were covered with clay. Although much of the clay had been used for making bricks, it still was on hand in such abundance that I had to wade in it over my ankles and had to burrow in with my hands, while from above the rain came down without mercy, but my gains were so plentiful that I did not let myself be diverted by all this. I had the pleasure, during the three days I spent here, of finding about 20 species of shells. Among each species was at least one which was as undamaged as if it had just now come out of the ocean which it once inhabited.

Friday, the 18th of October. All day the weather was such that in the literal meaning of the expression one would not like to chase a dog out of the house, but I just could not bring myself to stay at home; instead, right after breakfast I went on my way. Yesterday I had found a coral lump with a diameter of 1½ feet which contained on the inside the most beautiful crystals, some of which were more than 1 inch in length. But my hammer and chisel were too weak to do any good here; therefore I borrowed from an accommodating blacksmith one of his biggest hammers, took the same, besides my own instruments, under an open umbrella, and worked myself, with my load, through rain, storm, mud, and water to the mountain where my treasure lay, almost two English miles away. After many hard blows I split it into smaller pieces on which I could now work with my chisel and the smaller hammer. Very pleased, I withdrew toward noon, completely wet and so heavily burdened that I was forced to rest often, which was not exactly pleasant because I was exposed, below to the morass, and above to the cold rain. During the noon hour I changed into dry clothes, but alas, I had, as I left Jeffersonville, not counted on such continually wet weather and had brought along only one reserve suit; moreover, my little room could not be heated, and so I was, after lunch, forced to put on my wet clothes so as to have dry ones at my return in the evening. In the afternoon the weather was, perhaps, even worse, and toward evening, before I could reach the house, the storm broke my umbrella which had faithfully shielded me until now, and in the same moment, not yet satisfied, it tore off my hat, which fortunately was caught by a thistle bush and so was saved for me. If I have talked too much about the weather here, I think I have an excuse in that it has in the last few days exerted too much of an influence on me.

Saturday, the 19th of October. This morning I let it be my first concern to have my umbrella, so mistreated by the wind, repaired. To my joy, there was a good chance to do that because yesterday's strong storm had cleared the air so that the sky was free of clouds, and the nice weather lasted all day. Besides the repair of my umbrella, my main aim was to dry out my wet clothes, from the shoes up to the hat. This purpose I achieved by putting on my wet clothes and going in a hurry to the mountains, where I decided to finish my local work. My haste was for two reasons: first, I wanted to warm myself in my wet clothes by walking at a sharp pace, and second, I did not want to lose a minute of this beautiful day for my geological research. My efforts were crowned in every respect with full success, because the collection acquired here astonished even me through the diversity, beauty, and rarity of its objects. I could not think of counting, classifying, or even washing them, but packed everything as I had found it, to gain time.

This morning I collected a load of crystals, interwoven with corals, of the greatest beauty; I say a load because I was hardly able, even with great effort, to take it home at one time, and I had to rest every one or two hundred feet. In the afternoon I found a number of *Conchylia*, which were new to my collection, and some other nice objects.

Sunday, the 20th of October. This morning I went to church and in the afternoon I took a walk to the place where I had collected during the past week. As already mentioned, I had resolved not to collect here any more, but my walk convinced me that I had at least another half day's work to do here because I discovered a spot where two species of fossils are found of which I had, until now, only one piece each.

Bear Creek (four miles from the town of Dupont), Monday, the 21st of October. This morning at half-past five I left Madison by train for Dupont. The first two miles of the way were very dangerous to pass because it goes not only without interruption up a steep hill, but to get there a 100-foot-high embankment has been erected which is in such bad condition that in many places it has already started to settle and shows large cracks. When one has arrived safely at the end of this embankment, danger of another sort is revealed, because now one is at the so-called cut-through which I mentioned earlier, where at a height of 50 to 150 feet there hang suspended rocks of various sizes, which would seem ready at any moment to crash down on the railway. Some had fallen recently, and workers were busy clearing the debris away.

The morning was very cool, and the car open, so that we were very cold, and because up the hill the car was drawn by only four horses, the trip took rather long. Safely arrived on the mountain, we were taken to another car in which stood, to the delight of us all, a good hot iron stove around which all passengers took a seat, which worked out well, because the car was equipped to seat 48 persons, but there were only 14 passengers. From here it was only 11 miles to Dupont, where we arrived shortly.

Dupont is a five-year-old little town which has 6 houses and lies on the so-called Camp Creek; because there was no tavern, I stayed at the mail station, took a very good breakfast, and started afterwards right away on my explorations. In a short time I found on the banks of the creek some nice fossilized shells which were quite new to my collection. Soon I was joined by some curious boys, of whom one remarked that it would be a good thing if I went to see Mr. Wycoff,[2] who had a lot of fossils.

I learned that Mr. Wycoff was a cooper who lived ¾ miles from here, and I went at once on my way in the hope of learning from him something about the localities here. I found him to be a rather ordinary man but one who had a natural taste for collecting fossils and who was very obliging, since he not only gave me some of which he had duplicates, but he even offered to go with me and show me a place where fossilized corals were found of which he himself had examples and which were first-rate and quite new to me. At the place itself I found some of those corals which were very beautiful. After I drew Mr. Wycoff's attention to a species of shells which had brought me here particularly because of their beauty and rarity, he told me that those shells appeared especially at the Bear Creek, 4 miles from here, and there in fact mainly on the farm of his father. As indeed can be expected, I did not have to be told twice, but after I had hurriedly eaten lunch was on my way to Bear Creek, where I arrived after three o'clock and soon found my expectations surpassed, as I found just this evening a great number of lovely examples of the desired shells as well as a completely new, very beautiful specimen of encrinite. It was sundown, and therefore time to look for lodging for the night. Mr. Wycoff had told me in the morning that I could surely stay the night with his parents, and so I went to their residence, which consisted of a small log cabin to which an even smaller one had been added. Only two young people were at home, who told me that their father was absent but that they expected him soon, and said in answer to my question as to whether I could stay overnight that they could not say, for their father would have to decide about it on his return. So as not to lose any time, I continued my explorations although the sun had already gone down and I was very exhausted. I went not very far from the house, found a pile of rocks which had been thrown from the field to the side of the road, and to my surprise my first inquiring glance fell upon one of the biggest and most magnificent shells. I was not quite clear as to which family this splendid specimen belonged to, but it had much similarity with ammonites; it had, like them, chambers, all of which were coated with the most beautiful crystals.

It was almost dark now, so I went back to the house, where the old owner, in the meantime, had arrived; to my question as to whether I could spend the night with him I received a very vague answer which sounded more like a no than a yes, but he offered me a chair to sit down on. Under these circumstances I began to feel some uneasiness about my night's lodging and dinner; therefore I repeated my question with the addition that I wanted a definite answer because it would be high time for me to look for a hospitable lodging in case it was to be denied me here, and so I finally received the desired consent. Soon I was in conversation with my host, and he now became very friendly toward me.[3]

Madison, Tuesday the 22nd of October. My old host from Bear Creek was now so well disposed toward me that he accompanied me from daybreak until nine o'clock in the morning on my explorations and did everything he could to help me. After breakfast he even refused to take money for his hospitality and because I insisted upon it, he asked only approximately half of what was coming to him. In my explorations I was so lucky all morning that I decided not to eat lunch but to use the lunch time to work because between three and four o'clock in the afternoon I had to be in Dupont again so as not to miss the train. Among other valuable objects, I found pieces of two creatures which were entirely new to me. I was so heavily burdened that I made it only with great effort to Dupont by half-past three. Not only were my jacket, trouser, and vest pockets full, but I also carried a large kerchief, besides two single pieces, then my hammer and chisel, and, finally, the umbrella in my hands. Arrived in Dupont, I packed the whole collection in my travel sack, which was now so heavy that I could hardly lift it. Toward five o'clock came the train on which I departed. It was dark when we arrived on top of the hill which we reached yesterday with horse power. The engine was now taken away and the car in which we sat was let down the hill with such speed that the conductor continually had to brake with a device which was affixed for that purpose. When we were about halfway down the hill, a large pig came running quite calmly toward us in the middle of the tracks. With the help of the before-mentioned device the driver was able to keep the whole train stopped until one of the passengers got off and chased the animal away. Again safely arrived in Madison, I tried to make up at supper what I had missed at lunch, and soon went to bed.

Jeffersonville, Thursday, the 24th of October. Yesterday morning I paid a farewell visit to the place in Madison where I found so many beautiful objects, and I spent the afternoon packing them in a makeshift way because I had made up my mind to go back to Jeffersonville tonight. This evening the Whigs from Madison and vicinity had another festivity which consisted of a kind of torchlight parade and illumination of the town, but, as beautiful as the evening was, I could see the illumination only from afar because I was expecting the steamboat and could not leave the hotel, which lies quite close to the landing place. Some cannons were also fired, and the various fires burning on the hills surrounding the town on both sides of the river looked especially beautiful. Finally at eight o'clock came the steamboat *Tom Metcalfe*, on which I went to Louisville, where we arrived safely during the night. The boat was so full that I, with some twenty fellow passengers, was forced to sleep on the floor. In Louisville we heard the depressing news that the big boiler of the steamboat *Louise Walker*[4] had exploded yesterday evening and that the boat itself had been consumed by a fire in which many of the 106 persons on board had lost their lives and many had been severely burned or had suffered injuries to their limbs.

Saturday, the 26th of October. Yesterday and today it rained a lot, and I was busy from early morning until late evening with sorting, classifying, and packing the collection which I had brought from my last trips, a job which will take a few more days. Then I intend to travel directly to St. Louis because winter is coming. The Ohio is again rising, which will be felt in the fare of the steamboats, which previously was very high.

Sunday, the 27th of October. All day long it rained heavily, so I stayed in my rooms, except that in the morning I went to church although it was very muddy. The rainy weather gave me the advantage that I could carefully sort and pack my collection in peace. Also it can be assumed that, because of the rising of the Ohio, the travel fare will get still cheaper.

Monday, the 28th of October. The weather had cleared up, and so I went after lunch to New Albany, from where I returned soon, went on the steam ferry to Louisville, and from there on the train to Shippingsport, where there are many steamboats which do not want to pass through the canal which has been built to circumvent the Ohio falls between Louisville and Shippingsport and which I have mentioned previously. The local train is noteworthy in that one of the two cars that go here is drawn by a horse, the other by a small mule. The cars are covered and hold 12 persons each; nevertheless the stretch is covered at a continual trot. Arrived in Shippingsport, I crossed on the steam ferry again over to Indiana. Because Dr. Clapp was not in New Albany, I had to wait some time, and before I could finish my business so much time had passed that I became apprehensive that I would not reach Louisville in time to take the last steam ferry at sundown to Jeffersonville. In Shippingsport I found that the railroad mule had not yet arrived for the trip back. Therefore I had to rely on the speed of my own feet because the sun had already begun to go down, and I still had to walk more than three English miles to the ferry. I hurried very much and was indeed able for more than a mile to keep in step with a one-horse carriage in which sat a man who wanted to cross the Ohio at the same time I did. When he became aware that I was still close to him, he stopped and very courteously offered me a seat with him, which I of course accepted with thanks. In the course of the conversation I learned that this gentlemen was one of the representatives of the State of Kentucky; he took me almost to the ferry, and we parted from each other in the most friendly manner. Two minutes after I had boarded the ferry it crossed for the last time today to Indiana, and only thanks to the kindness of this friendly man I did not have to spend the night, against my will, in Louisville.

Tuesday, the 29th of October. The whole day long I worked in my room and did not go out. Dr. Clapp was with me this morning, as was another geologist this afternoon, to see the objects brought from my recent journey. Both gentlemen were not a little astonished at their beauty and at the variety of new discoveries which I had again made.

Wednesday, the 30th of October. One cannot imagine the excitement that exists all over North America because of the presidential election which will take place in a few days. Last night between twelve and one o'clock I was awakened by a serenade under the window next to mine. The singing was very pleasant, and it was accompanied by a violin and a flute; both instruments were played with tolerable skill, and the whole thing was very effective in the still, clear, moonlit night. This morning toward nine o'clock the windows shook from the thunder of cannons, and from the street echoed the drums beating and trumpets sounding from a wagon, drawn by four horses, which was followed by a long procession of riders and persons of both sexes in coaches. The procession went to Utica, 6 miles from here, where the Democrats gave a barbecue, and tonight, as every evening, the streets resounded with cries of hurrah and singing.

Sunday, the 3rd of November. Nothing happened in the last days except that yesterday toward evening I finished sorting and packing. I left my collection in Jeffersonville with a merchant for safekeeping because it is of too great importance and significance to expose it to the dangers of water travel in the winter. Yesterday afternoon I made a farewell visit along the falls of the Ohio, where I had experienced much quiet pleasure by finding so many beautiful relics of the primeval ocean.

✤ From St. Louis in Missouri to Bloomington in the Iowa Territory and back to St. Louis

ABOARD [1] THE Steamboat *Uncle Toby* 70 miles above St. Louis, Monday, November 10.[2] Before I left Jeffersonville, the Ohio had swollen for a short time to such a height that large steamboats could, without much danger, ride down its falls, where I had a few weeks earlier made excursions without getting my feet wet. This change had such an influence on the price for a trip from here to St. Louis that the fare, which had gone up to $20.00, was now down to $5.50. The trip, which usually took 8 to 10 days, can now be made in 3½ to 4 days.

Monday, November 4th, I said farewell to my friendly host and his family, and went bag and baggage to Louisville, where to my joy I found two steamboats destined for St. Louis. I made myself at home on the best one of them, called the *Lancete*. Only now I noticed that I had left my good hat in Jeffersonville, so I inquired how much time I had left before our departure for St. Louis and learned that it was an hour before our leaving, which wasn't much time. Still, I had a few minutes left when I arrived back at the boat with the left-behind hat. It was by now eleven o'clock, but, as it often happens, our departure was postponed until three o'clock in the afternoon. The weather was very beautiful; Indian Summer had just started, and it aroused really a very special feeling to see the steamboat glide over the foaming and swirling Ohio falls, which we passed without accident. We stayed only fifteen minutes on the spot where not long ago 106 unfortunates were blown up; [3] their mutilated bodies and torn-off limbs were still found from time to time and buried. A few days ago a head was found which was thought to be that of the captain.

It is strange that on the same day on which the *Louise Walker* blew up a second boat sank only two English miles from this spot, namely, the so-called *Emma*. The passengers of this boat escaped with their lives. It arouses a peculiar feeling indeed to ride over this spot, but with God's help we arrived happy and safe in St. Louis on Friday the 8th toward four o'clock in the afternoon.

On the morning of November 6th we were in Paducah,[4] in the state of Kentucky, where we stopped for an hour to load and unload goods, and on the afternoon of the same day, toward three o'clock, we landed in Cairo, a small town at the confluence of the Ohio and the Mississippi. Some years ago a company of Englishmen spent $3,000,000 to dam the shores of the Ohio and Mississippi which enclose Cairo, because they thought they could thus elevate the town to a great trading place. But because of the very unhealthy climate and some other reasons, the whole speculation came to nothing, and the little place affords a very depressing sight.[5] Here, at the mouth of the Ohio, a diving bell was busy salvaging goods of a wrecked and partly burned steamboat. The bell was fastened to a scaffolding which was held together by two big boats which were anchored here, and I was told that the person who went down often stayed more than an hour underwater. Six miles from St. Louis we again saw a diving bell occupied in getting the goods from the Mississippi from another sunken steamboat on which more than one hundred unfortunates had lost their lives. The weather during our entire trip was very nice, but we had so many passengers that many had to sleep on the floors and some on tables; I could hardly write, therefore, and so postponed it. Arrived in St. Louis, I met many familiar faces in the streets. The city itself had changed very much to its advantage, but I found the inhabitants had not. I used every moment to visit old acquaintances and gather news from them; thereby I found out that there was a very interesting site of fossils (of the bituminous coal formation) in the Iowa Territory, which had hardly or not at all been scientifically explored, and which promised a rich yield. My decision to go there was made at once, and, because I found that only a few more steamboats would make the trip before winter, I embarked without losing any time on the first, namely the *Uncle Toby,* and left St. Louis yesterday afternoon at three o'clock. We met two so-called snagboats, which dredge from the rivers the trees which are known in America as snags and which cause the sinking of so many boats. The weather was still very lovely. Our boat, to be sure, was one of the smaller ones, but, because we were only a few travelers, the journey was very comfortable at the start, and I wished that it had also ended that way.

Monday, the 11th of November. Nothing noteworthy happened during the first two days of our journey on the Upper Mississippi; it was only noticeable because of the weather that our course took us directly to the north, as the air seemed to foretell an imminent snowfall. Last night it was very dark, and I wondered how it was possible that we could travel all night long.

Because the Mississippi has at present a very low water level, there was fastened to each side of our steamboat a vessel of the size and shape of an Elbe barge.[6] The one on the right was called *Widow Wadman* and the one on the left *Corporal Tim;* consequently we traveled on a peculiar clover leaf. As we landed this morning to take on wood for fuel, one of our party caught an opossum in the cord wood, which appeared later at noon on the table as a delicacy.

Tuesday, the 12th of November. Our journey yesterday progressed without disturbance until an hour before sunset. At three o'clock that afternoon we arrived in Keokuk in Iowa, a place which got its name from a chief of the Sioux Indians who lives not far from here.[7] Here is the beginning of the so-called lower rapids of the Mississippi, where the river for about 12 miles opposes with great power the boats going upstream; adding to this disadvantage for us was a biting north wind. With heavy groans our old *Uncle* worked its way up about two miles, and because it rained we all sat around a big Franklin stove whose nice coal fire warmed us pleasantly. Suddenly there was a crash, and all passengers ran for the door to see what had happened because with the noise the heavy groans of old *Uncle Toby* also stopped and there was dead silence. But the situation was not as bad as one thought; merely something in the engine had broken and because the man attending it had stopped it immediately there was no further damage. The river had taken possession of the boat, however, and threatened to throw it against the rocks, which was prevented by quickly casting the anchor. Even under the most favorable conditions one could not think of going on before morning. All night long they worked to repair the damage to a certain extent. Nevertheless, we were told the depressing news this morning that the boat would have to go back to St. Louis because it was impossible to repair it sufficiently to continue upstream.

We could not even travel downstream because the wind was too strong. I therefore used the opportunity to unpack my hammer and stonecutter's chisel and made a little excursion along the bank which supplied me with some very interesting petrifactions. As the wind later subsided, the boat went back to Keokuk toward noon. Here I was on the bank, hoping that soon another steamboat would show up on which I could continue my journey. Keokuk lies directly at the foot of the rapids and a rocky hill and is an unprepossessing little place. I was assured from various quarters that it was very dangerous for travelers who carried money and valuables to stay overnight.

Aboard the Steamboat *Reveille*, Monday, the 13th of November. The small steamboat *Reveille* had fortunately soon followed the *Uncle Toby* but was so heavily loaded that it encountered great difficulties in trying to go through the Mississippi falls and was finally forced to return to Keokuk, where it was intended to switch most of the load over to a barge like the ones the *Uncle Toby* was traveling with. This enabled me to continue my journey this morning. Because of the heavy current it was slow going, and so I used this opportunity to walk ten English miles and make a geological examination of the rocky river banks on the Iowa side. To my not small delight, I found here a rich yield of beautiful petrified shells which were new to me, and also two species of corals. My greatest fear was only that the steamboat would travel too fast for me, but to my delight it made very slow progress, so that I could spend enough time on my explorations. Toward two o'clock in the afternoon I arrived, together with the *Reveille*, in the little town, where the goods, which had been brought on a small vehicle drawn by horses, were loaded on the steamboat again. Although very tired, I used the time it took to transfer the goods for another little excursion to a rocky brook where I did not find much. Toward evening our journey continued.

Thursday, the 14th of November. This morning at three o'clock the steamboat stopped suddenly, which woke me despite my weariness. I learned to my displeasure that the boat had run aground on a mudbank from which it could be freed only by unloading the goods. To bring about the unloading, the captain had sent right away for a vessel, which arrived after eight o'clock. The transfer itself took until four o'clock in the afternoon. Because the boat had to be completely lightened, all the passengers also had to be brought ashore. This would have pleased me if the river banks had been rocky and rich in petrifactions, but they were sandy, and a cold November wind blew mightily over them. Here was a small place with a general store where the passengers assembled themselves around the warm stove without buying a thing from the storekeeper, true to the customs of the Americans in the country, who sit often for hours in the store, talking, never even asking for one piece of merchandise, and then, without embarrassment, take their leave. Toward noon we were afloat again.

Friday, the 15th of November. Last night we passed by the Mormon city of Nauvoo; unfortunately I could not see anything because it was dark, but I got a good view of it yesterday afternoon although it was from far away. Several of our fellow travelers from the *Uncle Toby* joined us late at night when we stopped at the town of Burlington.[8] They had traveled to this place by carriage, which they rented when the boat was unable to go on, but they went from the frying pan into the fire because they had a very cold overland trip, much discomfort, and much expense, all of which we were saved. I had the additional advantage during this trip of adding substantially to my collection. The weather had changed again, so that today we had a beautiful day.

Bloomington [9] in the Iowa Territory, Saturday, the 16th of November. I finally arrived safely last evening after ten o'clock, after a journey from St. Louis which took more than six days and involved much more danger than usual. So, for example, last night we were only a few miles from Bloomington when we saw, sitting on a sandbank in the Mississippi, several hundred swans, in company with some Canadian geese, which craned their necks when we passed by, and for a long time could not decide to leave their resting place. So as to watch them more closely, I went up to the far end of our small *Reveille*, when one of my fellow travelers who had a good knowledge of engineering came up to me and told me that we were in great danger of an explosion of the right boiler. The machinery which was supposed to supply the boiler with water had failed to do so for a long time, and the captain had no intention of stopping until we reached Bloomington. To justify himself he drew attention to the force with which the right waterwheel was flung around. For about fifteen minutes we were in anxious expectation, and the few of us who knew of the danger, among them the pilot, went to the back part of the boat. But then the bell gave the signal for landing, as one can imagine to our not slight relief, and in a few minutes everything that had feet was on an island. The captain had at last begun to have some serious misgivings, for, after all, it was not only our lives but his also that were at stake. It took two men more than two hours to pump enough water into the almost-empty boiler, which was, after we arrived here, empty again. How glad I was when I could set foot on land. I went to the so-called Iowa House, a good inn which lies close to the river.

One can imagine that I was very anxious this morning to hear what I would find here, so right after breakfast I went to visit Justice of the Peace Mr. Porben, who had collected the first fossils here. Mr. Porben received me very cordially and told me that in a scientific respect I was well known to him although he had never met me personally. To my astonishment, I found in his collection many splendid and big pieces of well-preserved palmlike ferns indeed but only three pieces of leaves from this plant. It was also noteworthy that, when I visited Mr. Porben in his office in the courthouse, I discovered on the first stone of the step of the building a big piece of bark of a palm fern. The morning was spent in looking over the neighborhood, which did not give me much hope for great discoveries, but in the afternoon I came unexpectedly upon a spot where I found in sandstone a large number of the most magnificent primeval plants whose leaves were preserved very well. Among others, I discovered a fruit the size and shape of a walnut, which was beautiful and intact, like a flower bud with stem, almost ready to burst.

Sunday, the 17th of November. Today was not very productive as I did not trust the rainy weather for longer excursions; besides, I could not, on a Sunday, take the discovered items home without attracting attention, and so I had to leave them at the place of discovery until the next day. At noon I was invited to Mr. Porben's, whereby I was again robbed of several hours. During the time of my stay here I had obtained a whole fossilized splendid herbarium, which in addition contained many new objects. Bloomington is still a newly laid out town, only 6½ years old, with many tastefully built houses and approximately 1,000 inhabitants who, it seems, live in comfortable circumstances. This place lies on the bank of the Mississippi and is surrounded by hills which are of volcanic origin and sparsely covered with deciduous trees.

Monday, the 18th of November. Yesterday's discoveries were taken home today and packed.

Wednesday, the 20th of November. The collection which I assembled here consists of 400 examples of the most exquisite plant fossils which generally show so clearly one or more types of primeval plants that they can be classified without difficulties. Because a steamboat just now came down the river, I decided to go back to St. Louis at once.

On board the Steamboat *Iowa*, Thursday, the 21st of November. This morning I arrived again, with the Steamboat *Iowa*, safely at the rapids of the Mississippi, but the boat had such a heavy load that half of it had to be loaded into two large barges. This took four hours' time away, but shortly before that we landed, to my joy, in Nauvoo, the city of the Latter Day Saints, a place I had wished to see for a long time. This town, which was built only 5 years ago, lies in Illinois on the right bank [*sic*] 10 of the Mississippi on a plain which rises gently from the river. The river here has flat rocky banks and a good landing place. Some distance from the river is very fertile land. In the middle of the city, on its highest point, rise the walls of the not-yet-completed temple, which is 170 feet long and 80 feet wide and is of a very original architecture. The walls are strong and massive and decorated by some 30 columns which were worked by skilled sculptors. Very interesting in this temple is the place which is intended for the baptism of those who want to join this sect. This is an oblong water container which is borne by 12 life-size oxen. These oxen are real masterpieces of workmanship, and if viewed from a short distance they would be taken for live animals. At present they are painted white with oil paint, but they, like the water container they carry, will be gilded. Two of the oxen stand at each end of the container, and two pairs on each side, and between them there is a staircase which leads up to the container; from each side, stairs lead down to the water, where the person who is going to be baptized, like the Baptists, will be submerged by the preacher.

Not far from the temple is the foundation for an arsenal, for the Mormons are able, should they be in any danger, to take to the field with 3,000 men, a fact of which they are very proud. An apostle of the Mormons, a certain Mr. Pratt [11] whom I knew from Liverpool, showed me all the interesting sights. After I looked around town for a while, I went to the bank of the Mississippi to see whether I could discover anything in respect to geology. At first it looked as if I would not find anything, but suddenly I saw an undamaged petrified bone at my feet. At closer inspection I found it to be the thighbone of a *Labyrinthodon*, or primeval giant frog. Judging by the size of the bone, this frog must have weighed approximately 50 to 80 pounds and have had unusual muscle strength. But the really curious thing was that this piece was the first petrifaction of an animal belonging to the reptiles which was discovered in the layers situated under the bituminous layer. I say the first fragment because in a short time I found also a part of the tibia and a humerus of the same animal.

It was time now to get back to the steamboat, so I took, with one of my traveling companions, a small boat which got us to the other shore. After lunch we went on and in 1½ hours we were in Keokuk, where all the goods had to be loaded on again. During this time I conducted a small geological examination and returned toward evening with a few nice objects to the steamboat, which departed again at ten o'clock.

St. Louis, Saturday, the 23rd of November. When I awoke this morning I was safely in St. Louis. Because I expected mail from home, my first trip was to the post office, but nothing had come. My second concern was to find out whether my Dresden countryman and fellow traveler, Herr Advocat Ludewig, had arrived, but again nothing had been heard from him. With a third errand I was luckier. I wished to get the baptismal certificates of my children, which I had not taken when I left St. Louis four years ago. Fortunately, I found the same preacher who had baptized the children still here.[12] For the rest, St. Louis is lonely for me because all who were dear and beloved to me are missing.

Sunday, the 24th of November. I had already made reservations yesterday on the mail coach to Sulphur Springs in Jefferson County because St. Louis had lost all charm for me. In the morning I went to the church where my children had been baptized and listened to the preacher who had so often, in earlier years, edified me by his sincere and truly Christian sermons. Noon, afternoon, and evening I spent with various acquaintances.

✤ Sulphur-Springs and Herculaneum in Missouri

SULPHUR-SPRINGS, Monday, the 25th of November. This morning at three o'clock the mail coach awaited me at the door of the Virginia Hotel. The morning was very cold; the coach, though covered, very drafty because the window panes were broken and the sides were partly torn by tree branches which damaged the coach more or less on each trip, and it is surprising that nobody has thought so far of cutting them off. Besides me, only one other passenger was in the cold coach and we froze in competition until, toward seven o'clock, we found a warm room and a hot breakfast at the place where the horses were changed. But here I also learned that almost all my acquaintances in Sulphur-Springs had died, which put me in a considerable dilemma concerning a place to stay, because there was no hotel and I found it necessary to stay awhile. Toward nine o'clock I arrived here, and I asked right away at the nearest house whether I could stay a few days for money and kind words. The man of the house was absent just now, and his wife made some unnecessary excuses for not taking me in; one was that she had neither coffee nor sugar in the house, whereupon I answered that I would gladly be satisfied with milk. Enough; when the old housewife realized that I could not be rebuffed so easily, she let me have quarters, and soon we were good friends. I used the time until lunch and the afternoon for a small excursion. The host had not appeared for lunch, and when I came home in the evening I found him so drunk he could hardly stand up; nevertheless he received me cordially, and remembered me from the time of my first visit.

Tuesday, the 26th of November. Yesterday I found a fossilized creature, the genus of which I could not determine. The body is like that of a snake, more than two feet long, and the diameter is at the thickest part two inches. It is covered with a thick armor consisting of rings, which however leaves the belly and tail naked; the head, which lies deeper in the sandstone than the body, I have not yet been able to extricate. I also discovered on the rocky banks of the Mississippi two foot impressions that seemed to come from an aquatic animal of the size of an adult bear, to whose footprints they had some similarity—at least more than to the ones of the primeval gigantic frog called *Labyrinthodon* by Professor Owen. I was told that 16 to 18 English miles from here, at the so-called Rock Creek springs, there were similar footprints in the rocks; I resolved therefore to make an investigation of the spot.[1]

Thursday the 28th of November, 3 English miles north of Herculaneum. Yesterday after breakfast I went on foot to the springs of the Rock Creek. The way was so untrodden and there were so many forks that I had a hard time not to get lost; I could only move on with difficulty because more than twenty times I had to jump over the creek. Planks or bridges are out of the question here, and if it had not been so unusually dry, a person on foot could not have passed this way. Toward noon I had lost my way, but shortly I heard the sound of axes, and soon I was at a place where a party of farmers were building a log house, a style of construction which is quite common among the newcomers in the West. We had hardly exchanged a few words when three of them recognized me as the former owner of the museum in St. Louis. I could not stay but went right on after receiving information on which way to take. I was convinced that I would miss the midday meal because a restaurant, in this place, was unthinkable. But fortune decided otherwise. It was shortly after twelve o'clock noon when I came to a crossroad. While I was still undecided where to turn, I noticed in the distance a cooper's workshop. I went toward the small building, where I was received by large, half-starved, barking dogs, which finally, with difficulty, were calmed down by the man and woman of the house so that I could ask the way to Mr. Nool. I noticed unfortunately that I was not understood, so I repeated my question in German, and, to my surprise, received the answer that, if I could find the right way, I was only 2 English miles from my goal, but the name of the man for whom I had asked was not Nool but Noolen. Because I was afraid of losing my way, I asked the man urgently to lead me to the right path for a remuneration, which at first he refused, pretending business. But a few words from his wife brought forth a great change in him, and I was now invited into the house, which served at the same time as workshop, kitchen, bedroom, living room, and canary aviary, all of which was contained in a room 18 feet long and 16 feet wide. The canaries, however, were in a cage which took up one-eighth of the room, not counting the number of smaller cages in which were single birds. For lack of chairs I was offered a stool, and from the conversation I learned that my present host had previously been a forestry official in Germany with a salary of 1,100 florins; but he lost his post and was to take one at 250 florins, which moved him to emigrate to America with his wife and small daughter. After his arrival he went at once to the high West, bought a piece of land, and learned the trade of cooper, for which an obliging American neighbor taught him the necessary skill. He told me that he could now already make five flour barrels a day, while two years ago he was hardly able to make one a day. The hands of the poor man showed, however, that they still were not used to hard work because they were covered with scars.

I was cordially invited to lunch, which I accepted with thanks and without hesitation because I had no other prospect of getting a meal, and my walk had given me a healthy appetite. Right after lunch my host took me to the right road without accepting anything for his favors. His name was Goulard. After I had continued alone on my way for a short time, I came to a small house in which, according to the description of my obliging countryman, Mr. Noolen was supposed to live. But how great was my astonishment when, just now, I became aware that I had not only come to the wrong man, but had also gone in the opposite direction of my goal. The misunderstanding had happened through the mixup of the similar-sounding names, because, as I now learned, it was indeed Mr. Nool, and not Noolen, to whom I had to go to find the spot which I wanted to examine. Without prolonged reflection I set out at once on my way back, so as to get at least to the point where I had left the right path. But there were so many small byways that despite all caution I still got lost in the forest and did not arrive again until four o'clock at the house of my countryman, from which I had to walk another 7 English miles to reach my destination. It was therefore now too late to dare go any farther, and while I was considering what to do

I heard, not far from me, two men talking. When they had reached me I asked the older one whether there was a house nearby where I could get lodging for the night, whereupon he offered me his, which was near. This man happened to be the same cooper who had taught Mr. Goulard his trade. So as not to get lost again, I sketched the way on a piece of paper according to instructions of my host, and at daybreak I was again on my way. I had to use a very solitary and mountainous forest path and had trouble finding my way; in a few hours, however, I was at Mr. Nool's house, which lay in a beautiful fertile valley. But more disheartening news was waiting for me here, for I learned that there was no trace of footmarks in the rocks which I was looking for. Mr. Nool, who had lived here for forty years, and as a hunter had roamed far and wide through this region, did not know anything about it. He said, however, that he had heard from reliable persons that such footmarks had been seen near Herculaneum; to find out more about them I should ask a certain Mr. Herrington,[2] who lived three English miles from that town.

In five minutes I was on my way to Herculaneum; I had hardly walked two English miles when I was told that I had to go back one mile to get to the right way. This morning I had already walked six miles; two miles I went the wrong way, and from Mr. Nool's house to Mr. Herrington was eighteen miles; therefore I covered today twenty-six English miles over mountains and many small rivers. When in the evening I arrived very tired at Mr. Herrington's, I met there a doctor from Potosi who recognized me right away.

Friday, the 26th of November.[3] Mr. Herrington, whose house, as already mentioned, lies three English miles from Herculaneum, did not know anything about the footmarks which were supposed to be there, but he told me that a certain Mr. Giger in Herculaneum would know all the particulars if there really were any. Very annoyed by this, I made another try, however, as long as I was now here, and went to Herculaneum but found nothing there too. Still, Mr. Giger called my attention to a recently built chimney in which showed, indeed, the larger part of the footmark of an animal the size of a bear. Now I had at least a chance to investigate further the place where the stones had come from, and soon I had the pleasure of finding the footprints of two animals and, besides that, a number of beautiful new petrifactions.[4]

Saturday, the 30th of November. Because I have mentioned Herculaneum only casually, I want to add that this town really has some similarity to the one whose name it bears. That is, that the new Herculaneum as well as the old one is buried—the first one by volcanic forces, the second by the force of the Mississippi, which has thrown a large sandbank in front of the otherwise-flourishing new Herculaneum.[5] This of course does not cover the houses but has cut off the remaining inhabitants completely from business endeavors. Only six families live here still; all the other houses stand empty and deserted; everywhere one sees remains of old business firms and broken windows. At the last great flood of the Mississippi some of these houses were under water up to the third floor. One can still clearly see the flood marks. The whole thing looks very sad.

This morning I found some nice petrifactions, among others the complete upper arm bone of a giant frog. In the afternoon I started on my way back to Sulphur-Springs and I passed a house in which recently a German cobbler and his young wife had been murdered by a Negro who had hoped to find a lot of money.[6] In this he was misled because he found only $7.00. First he killed the man, who was working outside the house, and then the unfortunate woman, who carried a 1½-year-old child in her arms and was expecting soon. The child also received a wound from the three blows which the murderer inflicted on the woman with a meat cleaver. The cries of the child attracted the attention of a passing farmer, and he pursued the murderer, whom he identified through bloodstains on his shoes. He was handed over to the law but was in jail only for a short time when about 300 men and women of the neighborhood collected in a mob, demanding immediate punishment of the criminal, and, despite all remonstrances by the few law officials, broke down the jail door and took possession of the murderer. First, they disagreed as to whether he should be burned alive or hanged, because a large part thought the punishment of hanging too mild, but they decided for the latter, which was done without delay.

Sunday, the 1st of December. Today I decided to embark on a new journey, because during my absence my good old landlady had heard that 40 English miles from here there was a place where various footprints of humans and animals were to be found in the rocks. This information came from such a credible man that I did not want to spare the expense and troubles of the trip.

Wednesday, the 4th of December. Last Monday morning I took the stagecoach which leaves every two days from St. Louis for Potosi. At noon I was already in Herculaneum, where a nice wild turkey was eaten for lunch. Toward four o'clock I came to Hillsboro,[7] a small town where, as in Herculaneum, the horses were changed, whereby I had time to see a small mineral and petrifaction collection belonging to a certain Mr. Mathews. I was especially interested in a footprint pressed in rock, of the size of a 14-year-old human, which had an extraordinary resemblance to one that would be left by a foot wearing an Indian shoe. This impression was found ¾ mile from here when the rocks were broken away. It was after six o'clock in the evening when I arrived at a farm 1½ miles from Mammoth-Lead [8] and 17 miles from Potosi. The next morning I went on foot to the new mine, which received the name Mammoth-Lead because of its rich yield of lead. Here I looked up the man who was supposed to show me the tracks in the rock, and I found him to be the owner of the mine, already an acquaintance; he lent me his horse for the continuation of the journey, which I accepted with thanks because there were various rivers without bridges to pass. After I had pursued my way in this manner up hill and down dale I came to a mountain ridge sparsely covered with red cedars and low deciduous trees, on top of which showed the first traces of footprints. The greater part of them, however, had been carved in the rock by hand like some other figures, for example, a three-pronged implement, a sun, and so forth. Several, though, were not artificial but had been caused presumably by creatures who walked there when the stone was still a soft substance. Between two juvenile footprints it was difficult to differentiate whether they were artificial or real.[9] I rode now to a house to find out whether there was someone who would chisel out the footprints I wanted to take with me. The owner of the house was absent, and a young man who looked after his business said in a few words that the owner of the house would under no circumstances allow these footprints to be removed. Then I rode sullenly back to the mine, where I collected still a few nice minerals so as to have at least something from this trip.

Thursday, the 5th of December. This morning I went back to Sulphur-Springs with the mail. The coach in which I came here was bad enough, but at least it had springs so that one could stay in one's seat when it went over tree trunks and medium big rocks in a trot, but now

I had a coach in which despite all efforts I could not stay in my seat. I therefore took a seat on the floor between a suitcase on which the coachman sat and the seat intended for passengers, bracing myself firmly with my feet and back, and holding on with my hands; fortunately nothing hindered me from doing that as the coachman and I were alone in the coach. At the first station we had breakfast together at a table, and it was said that from now on I would be accompanied by a lady, whom the proprietor asked the driver to wake because she was staying at another hostelry. I began to feel anxious when I thought of how we would sit, or rather how we would keep our seats, when the driver came back with the news that the lady had changed her mind and would not travel today, and so I arrived safely here this afternoon.

Saturday, the 7th of December. Some time ago I had ordered a few tools from the local blacksmith, which I needed for breaking out a rock plate which was precious to me, but the blacksmith was a constantly drunk and lazy man, so that only yesterday evening, after all sorts of maneuvers, was I able to get the tools. This morning I went with my old host 4 English miles to extricate the stone plate, which in geological respect is of the highest importance. It shows two footprints of a creature whose foot was similar to that of an almost fully-grown human, but the foot had been much wider in front and had consisted of five toes and one thumb. The distance of the right from the left footprint is the same as that of a walking man. On the same plate are the left and right footprint of a bird the size of the largest still-living herons.[10] We had hardly worked an hour when our tools were useless. Fortunately the house of Captain Waters [11] was nearby, and through the kindness of his wife we received some tools with which we finished our work. A few years ago I lay very sick in the house of this dear family and was treated by Mrs. Waters as kindly as only by one of my closest friends.

Sunday, the 8th of December. Yesterday I noticed a place where there were beautiful fossilized shells which until now I have found only very seldom here, so this morning I made an excursion to the spot which was more rewarding than I had expected because I found enough shells to determine with certainty the geological age of the rocks which for a length of 200 English miles extend along the banks of the Mississippi and also stretch into the interior of the state. Until now this rock formation was judged by all who had inspected it to be mountain limestone, but already at the time when I made my first examination here this seemed to me to be an error. I preferred to be quiet about my conjecture until I could substantiate it. Now the time has come because I have all the necessary evidence in my hands to refute the error. The first theory was off by one geological period. The local rock formation is old red sandstone, and only above this formation would the mountain limestone be found—if it were really here, which is not the case.

Monday, the 9th of December. This morning I rode with my old host to a point near Herculaneum where I thought I had seen the footprint of a huge bird. We were equipped with tools to break out the rock plate which contained the print and to carry it on the wagon to my present lodging. It was one of the coldest days we have had this winter; we were very cold and were forced to stop twice to warm ourselves, and, moreover, the road was very bad. Finally, between twelve and one o'clock we arrived at our objective, where we made a fire and went to work. However, to my annoyance I found that the water which probably had washed for thousands of years over the footprint had defaced it so that many doubts could arise about its authenticity; I therefore decided to give up the work to save myself wasted effort and considerable cost. Thus we went back home, where we arrived toward evening.[12]

Tuesday, the 10th of December. This morning I had sent a man with an oxcart to the place of the earlier mentioned footprints to take them to the house of my old host. On my way there I found some more very nice petrifactions and I arrived there at the same time as the wagon. Fortunately two farmers, just passing, helped us load the stone plates; in a few hours we had them at the place from which I was going to take them to St. Louis.

✤ Golconda in Illinois, Smithland in Kentucky, Natchez, New Orleans in Louisiana

ST. LOUIS, Wednesday, the 11th of December. So I have finished another round and arrived again, with God's help, in St. Louis, where in earlier years I have experienced so many joys and sorrows. Last evening and this morning I had my collection, as well as the stone plate with the footprints, loaded on a wagon with two horses and toward evening arrived here safely and handed the plate over to a stone worker who was to make it a little smaller and lighter.

Steamboat *Palestine*, Monday, the 16th of December. Yesterday evening I worked late into the night to pack my collection so as to be able to go on the steamboat *Palestine* to Cairo this morning. But my hurry was in vain because the departure did not take place. Winter showed itself in all its might. The Mississippi was so full of ice floes that I started to worry about the steamboat trip; in the evening it also snowed very heavily. This noon the steamboat finally departed; but 35 English miles below St. Louis we berthed again because our captain did not dare, despite the most beautiful moonlight, to travel during the night.

Tuesday, the 17th of December. Not until nine o'clock this morning did the captain again continue the journey. The ice was so thick on the river that one could see only a little water and sometimes none at all. At noon we noticed two steamboats ahead of us; the one which had left St. Louis two days before us was grounded in the middle of the Mississippi on a sandbar and was in a very critical situation, particularly in great danger of being crushed by the ice; the second one worked its way slowly up river through the ice. At the sight of this, our captain became alarmed and landed on the shore. The steamboat traveling up the river did not know why we landed and out of concern followed our example. After lunch I went with several other passengers of the *Palestine* to visit the ones of the other boat, which lay at a distance of more than two English miles away. In the vicinity of this boat I found some magnificent receptaculars. Unfortunately I did not have my hammer and chisel with me; I went as fast as possible back to the *Palestine* and fetched both. Because I did not know for sure whether the boat would dare to enter the river again this evening, I could in my haste obtain only a few examples of the above-named fossils.

Wednesday, the 18th of December. Last night the weather changed so that today it froze only a little or not at all. Already during the night I had noticed this change, favorable in more than one respect, when I woke up and found to my astonishment that I did not shake with cold but felt rather warm. Our mattresses have only a very narrow quilt and a sheet for cover, which for the summer is good and sufficient but not for the cold time of year. Because of the mild weather the captain had taken courage again and, to the joy of all passengers, started up this morning at about eight o'clock. Shortly we came to the boat lying on the sandbar, which was now frozen in the ice. Its captain tried to persuade ours to take his cargo to shore, but he refused for reasons of self-preservation, and now all the passengers of that boat streamed onto ours leaving it to its fate. Among others were a number of milk cows and fat pigs. The whole misfortune had been caused by a drunken pilot.

Toward twelve o'clock we were forced to stop for a short time. Because of the ice we could not get to the land, so a long spar was laid from the boat over the ice. With several passengers I climbed on shore to see how things were concerning fossils. Since I could not find anything, I dared to go a little farther with another companion, but soon I began to get worried because my companion had seen somebody hurrying toward our boat, which indicated that the same would probably depart. We therefore ran with all our might to our floating home. A half minute later we would have been put into an exceedingly embarrassing situation, for already the warning bell was silent, and the moment we reached the spar we heard the bell at the sound of which the big engine is started. A man who became frightened when I did not walk fast enough for him almost lost his life as a result for he wanted to pass me and in so doing ran a few steps beyond the plank. Suddenly the ice broke under him, and only by grabbing for the plank did he save himself, for the next moment the current would have pulled him under the ice, and he would have been irretrievably lost. Later we passed through a strait narrow pass which was so full of ice floes that one could not see the water, and it was only thanks to a change in the weather that we could go on.

Thursday, the 19th of December. In the last 24 hours our situation has again changed, but alas! not in our favor. During the night, when we as usual lay still, the steamboat *Alleghany*, which had left St. Louis 24 hours behind us, passed us. The *Palestine* made all possible efforts to catch up with it, and this morning toward eleven o'clock it succeeded, but already both boats were so surrounded by ice floes that we could only drift with them, and within half an hour both were firmly ice-bound.[1] In the afternoon three people dared, with the help of a board which they pushed from one floe to another, to go to the distant shore. Their purpose was to find out how things were farther down the river. After several hours they returned without any comforting news. During their absence several attempts were made by our captain to push us through the ice, but all in vain. Among other things, all persons were called to the uppermost deck where we, at a given signal, all at once ran from one side to the other, whereby the boat was put into a rocking movement, which the steam engine was made to use all possible power, but this also was in vain.

Golconda, Thursday, the 26th of December. The night of the 18th to the 19th of December was a terrible one for us. As already mentioned, we were locked in drift ice and all of us were anxious as to what our fate would be. At one o'clock at night a dull rumble, like distant thunder, was heard. This ceased but was repeated toward five o'clock, when it was joined from time to time by a certain cracking noise. Now the whole frightful mass of ice moved; our boat trembled and moved slowly forward straight toward the other steamboat, which lay at a right angle before us, and which we would have shattered if the mighty force of the ice had not also dragged it along and stopped us at the moment of the greatest danger. But as soon as we stood still again the ice floes threatened to crush the boat. The sight was indescribably dreadful for there was no thought of escape unless the ice closed up again, which fortunately happened shortly. All anxiously awaited the dawning day, and when it finally came everybody tried to save himself before the ice broke loose a second time. The first who walked over the ice to shore were passengers of our unlucky neighboring boat, which had already sprung a leak. Some of our party followed soon. The worst now was to get the baggage on shore, and high prices were paid to the black servants who were brave enough to carry suitcases to the shore and from there two miles over a rocky hill to a place where wagons could travel. A part of the travelers went to the Missouri side, which was 1¼ English miles from the boat, while the rest had only to walk ¼ mile to the Illinois side. Among the latter was a gentleman with a four-year-old daughter whom he carried with great difficulty. He and his companions had to wade for a long stretch in water one-and-a-half feet deep which had overflowed the ice. I with several others stayed on the boat until afternoon. Fortunately the news of our situation had spread in the neighborhood and moved some speculative Americans to clear a sort of road through the forest with three wagons to take our belongings to a place from which they could be transported farther on. Everybody took advantage of this opportunity, but when I, who had taken a shorter way, arrived in the evening on the spot where we awaited our suitcases, somebody congratulated me that I had saved not only my life but also my things. This weighed heavily upon me because until now I had still hoped that in time the boat and the load would be saved, but now I heard from all sides that there was little or no hope for that. Three boxes and a suitcase which contained all my geological treasures, collected with so much trouble in Iowa and Missouri, were still on the boat. I decided therefore to urge the oxcart man, who just now arrived safely with the first load of suitcases, to try salvaging my things the next morning. The transaction with this man was soon concluded, and I seated myself on the oxcart with which I still had to travel this night at a snail's pace 10 English miles to be near the boat in the morning. Toward nine o'clock we reached the house of an acquaintance of my driver. Here we had planned to eat dinner but found everybody in deep slumber; without ceremony my companion woke the lady of the house with the statement that if she knew what we wanted she surely would not stay in bed any longer, and, indeed, in a short time we had roast pork, cornbread, dried apples, and coffee.

The country through which we traveled was most romantic, especially in the magnificent light of the full moon. On one side were 200- to 300-foot-high cliffs covered with various vines and trees; several brooks rushed from rock fissures over big icicles; at three places I noticed entrances to caves. The opposite side of the road was bordered by a primeval forest with trees of a size and strength of which one has no idea in Germany. To give the whole an even more romantic atmosphere, from time to time the howling of hungry wolves could be heard, sometimes close, sometimes far, and through the rustle of the wind in the tops of the gigantic trees I heard at short intervals the wailful cries of several horned owls, which indeed harmonized not badly with the adventurous night journey on the oxcart.

During the night, toward twelve o'clock, we arrived at the home of my driver, which was one of the saddest I have ever seen. The man was one of the large number who had lost all their goods and chattels in the last great flood of the Mississippi and had saved only their lives, which many later lost to the virulent fever. Even the land of this unfortunate one, like that of many thousands, was made forever completely unfit by sand several feet high, and now he lived on strange deserted property. Neither door nor window had the small wooden house, and many holes in it were so big that a man could easily pass through them. All of it threatened to collapse with the next strong wind. An old, small table, a broken chair, and two wretched beds were all the furniture. The woman of the house looked more like a ghost than a living being, and in the bed, which she had left at our arrival, lay a loudly wailing twelve-year-old boy who had recently survived the so-called winter fever which had snatched away so many, and now suffered terribly from a painful fever blister. The poor woman had kept coffee warm for her husband, which he shared with me despite my reluctance. Now a dirty boy of ten, a girl of nine, and one more boy of six were shaken from their sleep; the two latter ones were told to sleep with their sick brother and the mother in one bed, while the dirtiest boy was to share a bed with his father and me, so that the man and I lay on the upper part of the bed and the boy on the lower part between us. Breakfast was prepared in an old broken saucepan, the only utensil in the house. Although I tried to overcome my disgust, I could hardly eat anything, but because I did not know when and where I would eat again, it had to be done. With uneasy anticipation I now approached the place where the unlucky boat with my things was lying, but to my great joy I found that the ice had not moved during the past night, and everything was still in order. I was now looking for somebody who would, with the help of my driver, take the boxes on land, but in vain. There was no time to lose; it was thawing, and every minute the ice became more unsafe. I gathered all my strength together and with the help of my driver I carried on a barrow first the suitcase and then the two lighter boxes to shore, but the last box both of us could not move, and all efforts to get people to help were for a time in vain, until finally two mulattoes let themselves be persuaded to carry the box on land for the price of $1.50. The ice had already become so unsound that those people broke through it three times with one foot, and half an hour later it would have been impossible. How glad I was to have brought everything on shore only he can imagine who knows how irreparable the loss would have been for science and for me.

All the rescued ship's party who had their luggage transported by wagons had stayed at the place where it was unloaded and were hoping that one of the five steamboats lying close by, and unable to go up the river because of the ice, would take them along, even for a high price, to the mouth of the Ohio. But this hope was also in vain because, as already mentioned, thaw had set in and they were intending soon to continue their trip to St. Louis. I do not know what became of all the ladies and gentlemen, but they were in a very alarming situation when I left the place, for their number was more than 70 and there was almost nothing to eat any more in the place. As for me, I had parted from the rest with 3 gentlemen and one lady, and we were lucky enough to find lodging in a good country house. My boxes and suitcase with the fossils I left with a reliable man for safekeeping until the coming spring. We rented a wagon drawn by four oxen which took us with our luggage 5 English miles. Here we had to cross the river on a small ferry and try to find a new conveyance. One of our party started off for this while the rest of us camped around a fire. After two hours our friend came back with a rented wagon drawn by two horses. Our intention was to reach the nearest settlement on the Ohio and there again embark, each one to his own destination. Without particular adventures we arrived last Sunday at a house where we could spend the night, after we had several times asked in vain for quarters for the night. On Monday before breakfast we continued our travels, but because our driver had only to take us 4 English miles more as agreed, and his cart threatened to break apart at any minute, the gentleman who had helped us once before tried again to procure a wagon for the 60 miles which we still had to cover. Monday toward ten o'clock in the morning we reached the spot where our driver was to leave us; one of our party went into a house on the road to order breakfast for all of us but came back with discouraging news for our hungry stomachs; since the people in the house had been slaughtering pigs they were too busy to fix breakfast for anyone. I was dissatisfied with this excuse and went into the house and asked permission to warm myself at the fire; after I received it I made the remark that it would indeed be wrong that hungry travelers who wanted to pay for their breakfast could not get anything. This got action, and I suppose that the word *"pay"* moved the man to declare that he would serve us. We received indeed as good a breakfast as we could have expected. After eleven o'clock our friend arrived with a good wagon and two strong horses. We covered this day 34 more English miles, because no lodgings for the night could be found any sooner, and arrived in the evening in the small town of Vienna. Here we had to knock with all our might on the doors to get a look at the innkeeper or anybody else. When finally a door opened, the proprietor came rushing out with his whole household, very astonished to see a wagon with six strangers. To our question of how it happened that he did not pay more attention to the arrival of strangers, he said that often in a month not a single one came to him, although he was the only innkeeper in Vienna.

Last Tuesday, the day before the first Christmas holiday,[2] we still had to do 26 English miles, but, because the road was extremely hilly and the driver had to exert his horses too much the day before, we arrived in Golconda only toward evening. During our stay here many steamboats passed by, but none could be induced to land because, as I heard, it was so difficult to do. I would have traveled on to Smithland to get on a steamboat again, but I found here, quite contrary to my expectation, an old rock formation in which I soon discovered five species of fossils completely unknown to me. At the same time I learned that there was a very interesting formation 12 to 15 English miles from here.

Golconda,[3] Sunday, the 29th of December. Wonderful and inexplicable are the ways of our all-gracious heavenly Father; this has been proved again to me in the last few weeks. Despite all the trouble I went to to leave St. Louis before the ice in the Mississippi increased, I could only get away on the *Palestine,* from which I saved myself and my things at the apparent risk of my life. I made all possible but vain attempts not to come to this place, the last one on the morning when I left Vienna (in Illinois) with my small traveling party. There we were advised strongly against going to Golconda because, as they correctly pointed out, we would not get a steamboat to continue the journey, but the majority of our small party paid no heed to the warnings, and finally it went so far that we drew lots to decide whether to go here or to another landing place on the Ohio, and I myself, who had absolutely no intention of coming here, chose, guided by a strange feeling, a lot which made me change this decision. And it is this circumstance which, as I firmly believe, decided a great part of my and my family's welfare. As I already remarked, I found here at Golconda a place with magnificent fossils totally unknown to me, and I intended to ride in the company of another traveler, named Graham,[4] to another spot to make new discoveries there. We tried to obtain horses for this, which entailed many difficulties, but we succeeded to the extent that I got a horse with saddle and bridle while my friend had to be content with a gray mule; both animals, however, were blind in one eye. The day before yesterday we wanted to leave here very early, but it was ten o'clock in the morning before we could get together the horse, mule, saddles, bridles, etc. After twelve o'clock we arrived at our destination, and soon I had the pleasure of reaching an area where again I found a number of nice unknown fossils. But how great was my astonishment when the man with whom we were lodging led us to a place which I almost want to call a lead mountain; at least I had never seen a rock formation where lead lies so abundantly and pure in the open. At many spots holes like wells had been made, and I did not see one which was not surrounded more or less by levels of lead. Some of the holes were partly filled with water, which had driven the workers away, because they could not afford a machine with which to get rid of the water. During our inspection of the lead mines and collecting of fossils evening had come, and so next morning we rode with our host to a steep hill, 2 English miles from our quarters, where, to my astonishment, the scene of the past day repeated itself. After we had inspected the place thoroughly our host showed us still a third spot which, still intact, also promised a very rich yield. The two aforementioned places (each 40 acres) belonged half to our host and half to a man in Missouri. Before we rode back to Golconda in the evening we had bought the part of the land which belonged to our host for a low price and were now owners of one of the richest lead mines in the United States.

Monday, the 30th of December. This morning my companion and I started out on horseback for a lead mine, 11 English miles from Golconda, of which we had heard a great deal. The way was difficult to find, and, instead of the approximately 22 English miles it should have taken to get there and back if we had found the right way, we had to cover approximately 32. Since we found on the way a landing place on the Ohio which is about 8 English miles from our mine and is very suitable for shipping the metal, we purchased here temporarily 8 to 10 acres of land. Unfortunately we found that the lead mine we visited was much less productive than it had been described to us, as it stood no comparison with ours.

Tuesday, the 31st of December. Last year my affairs forced me to spend the last day of the year far from my family in Berlin; this year, too, I am separated from them and I write these lines far from all human society; nevertheless I am not alone because the all-benevolent God who is always near in joy and sorrow, in the noise of the world and in solitude, leads my thoughts to my dear ones in the distant Fatherland.

Wednesday, the 1st of January 1845. I spent New Year's day as quietly and alone as last night. For several days now the weather has been as beautiful and warm as it is with us at the beginning of May; the birds sing in the forest and I spent a long time in an unheated room. Under almost every stone which I turned in my search today I found living insects, and a few days ago a butterfly sat on my jacket.

Thursday, the 2nd of January. This morning Mr. Graham and I rode to the spot, which we found suitable for shipping the metal, for the purpose of looking it over carefully and, if it met our expectations, of buying it. As it happened, the owner was not at home when we arrived, so we had a chance to inspect everything again undisturbed. On shore lay an unsecured skiff, and soon we found a pole and two oars which we appropriated without ceremony for a time, so as to test the depth of the river here. We were convinced to our satisfaction that even at the lowest water level of the Ohio the biggest steamboat could comfortably land here, and also that the river bottom and the gently rising shore consisted of sandstone. The shore offered one of the finest views, so that on a moderately clear day, as it was today, one can see ten English miles down the river and several miles up. The landing place in Golconda is so bad that now almost no steamboat dares to land there; accordingly,

Golconda would come back significantly if a new town were to be built five miles farther up the Ohio. Toward evening we had closed the deal with the owner.

Smithland,[5] Thursday, the 9th of January. Today I was ready to continue my trip to the South. I had taken my things to a grocery store near the shore of the Ohio and made an agreement with a man who owned a boat to take me and my luggage to the first steamboat which came down the Ohio. The sun had not yet gone down when such a vessel appeared in the distance. By waving my handkerchief from the little boat I signaled that I wanted to travel with them and, as a sign that they had understood, I heard the bell of the boat ring. The boat, called the *Emma*, was very small, and its destination was Nashville in Tennessee. Therefore I could only use it for 17 English miles because, from Smithland on, it took a different direction from my travel plans.

Saturday, the 11th of January. After I had waited yesterday almost 24 hours in vain for a New Orleans steamboat, the steamboat *Grace-Darling* took me on board after sundown, and now I was going down the Mississippi; in 24 hours we had already covered 200 English miles and so I have come a good bit closer to my goal, the city of Natchez in Mississippi.

Monday, the 13th of January. This morning we met a boat with a diving bell; it was loaded with many barrels of brandy and other goods which had been recovered from a sunken steamboat. This boat was on its way to a second wreck. Shortly before lunch we landed at an island to load wood. It happened that the shores of the river were so situated that we could easily get to land. The weather was very nice and warm. I noticed clearly that I had moved toward the South because I found not only several beautiful plants in full growth but also one with blossoms and green seeds. After lunch we came to the first trees which were covered with so-called Spanish moss (a species of moss which sometimes grows to a length of many feet and gives the trees from which it hangs quite a peculiar appearance). It is dried and used much for filling mattresses. It looks very much like horsehair.

Tuesday, the 14th of January. Today the weather was as warm as in Dresden at the end of May or the beginning of June. In the afternoon we arrived in the city of Vicksburg, where we had time to go on land a little and look at some gardens in which everything had the appearance of advanced spring. Among other beautiful plants I noticed lovely, rambling roses decked out in spring green which not only densely covered several arbors but also were full of buds which promised to break open very soon. What a pleasant sight for a northerner! The city of Vicksburg lies on a high hill which, where there are no houses, is covered with luxuriant grass. On this grass were grazing horses and cows, and, to make the sight even more romantic, a flock of more than 100 dappled goats frolicked there.

Wednesday, the 15th of January. This morning toward five o'clock we arrived in Natchez, where I had intended to spend some time, but yesterday late in the evening I decided to continue my journey to Alabama without delay because, as I heard, extremely little rain had fallen in the South, and the smaller rivers and creeks had therefore a very low-water level, which could be only too fortunate for my explorations. Toward noon we took on 25 cords of firewood, whereby we found time to go ashore. From a short conversation with the farmer where we loaded the wood I learned that he had sowed oats this morning and that his recently planted potatoes had already sprouted.

Thursday, the 16th of January. This evening I arrived well and safe in New Orleans, and because I intend to go on early tomorrow, I am extremely pleased that I can stay on board the boat until then. This evening the air is very warm and sultry. I have spent almost all day looking at the splendid plantations which lie in an almost continuous line on both sides of the river. I especially enjoyed the wonderful orange trees in the gardens which surround the very elegant houses of the Louisiana planters. Many of the trees were so heavily laden with fruit that the branches bent low. The streets in New Orleans are very dirty, so in the evening I returned very soon to the boat from a small excursion to the city.

Mobile, Sunday, the 19th of January. I had just finished writing the last lines in my diary in New Orleans when a terrible rain started which lasted until Friday morning; also a wind came from the north, and the air grew very damp and cold. I delivered a letter of introduction which I had received in St. Louis for a local commercial house, found a very friendly reception, but could not stay long because my things had to be taken to the train which goes from New Orleans to Lake Ponchartrain. With a one-horse carriage which cost exactly three times as much as in other places, I arrived toward eleven o'clock at the railroad, a distance of 1½ English miles from the steamboat. Here everybody had to help himself, because when I asked, while paying, where to put my suitcase and the rest of my things I was told just to put them in any car. After three-quarters of an hour we reached, without particular adventures, the mentioned lake; but here the trouble with my luggage started, because the train stopped ¼ English mile from the place where the steamboat was moored. It was raining and nobody was here who could have taken my things to the boat. I was therefore forced to carry one piece after another myself, but finally I found a worker who helped me. Because it was noon and the steamboat destined for Mobile in Alabama did not leave until after three o'clock, I went to a hotel which was situated like a summer palace in a beautiful garden close to the shore of the lake. Here I ate my noon meal and afterwards took a walk in the garden where among other plants five varieties of roses were in beautiful bloom. Many oleanders as well as a large aloe had buds; the tidy paths were covered with sea shells, for gravel can hardly be obtained here; it has to be brought here by ships, often from far away. It was truly a delightful pleasure to take a walk in such a garden on the 17th of January. From the flower garden I made a trip to the vegetable garden; here, too, everything showed in luxuriant green—lettuce, cabbage, some recently planted, some already developed into the most beautiful and perfect heads, onions, radishes, and so on.

Anchor was dropped toward midnight because, as we heard later, our pilot got lost, and not until Saturday at daybreak did the journey continue. Had we not lost our way we would have avoided the Gulf of Mexico, but now we had to pass through it for a stretch, and the worst was to get into it, for we were between a number of oyster beds on which the waves sprayed in terrible breakers into the air. The local oysters are indeed very good (as I found in New Orleans where the captain had treated us to an excellent oyster dinner before we left the boat), but as there can be too much of a good thing, so it was here; for while several dozen oysters would have been very welcome if we had had them on the table, if the great mass of them had got under the boat it would have caused us to sink. Still we passed safely, after a few vain tries, and landed at noon in Mobile.

✤ Claiborne in Alabama

CLAIBORNE,[1] Monday, the 20th of January. In Mobile I found at once a steamboat which, toward evening, would go up the Alabama. I boarded it without delay, especially because it was close to lunchtime and I therefore could avoid the very expensive restaurants. Uncertain when our departure would take place, I did not venture far from the boat, mainly because I did not find anything in town which could have drawn my special interest. To my displeasure, our departure was delayed until dark. Because the 144 English miles from Mobile to Claiborne were covered in 12 hours, I had no chance to observe closely the banks of the Alabama until we were close to landing. I was very glad to have reached Claiborne safely because four steamboats which left Mobile simultaneously going in the same direction kept up a real race the whole night through, whereby we were put in no small danger; a few times during the night I was afraid the boilers were going to explode.

Claiborne is a very romantically situated town of about 400 inhabitants; the streets are straight and wide. The sidewalks (which like the streets are not paved but kept clean) are planted with so-called flowering ash. The town lies directly on the Alabama, but the banks of the river here are 200 to 300 feet high and almost perpendicular. At the steamboat landing is a wooden staircase of 345 steps which leads to town. To transport the luggage (as well as the merchant goods) to town, there is a kind of train which runs parallel with the staircase. The car is fastened to a thick cable and is pulled up by a large wheel which is driven by several mules. After I had climbed the steps I was almost out of breath, and I awaited the arrival of my luggage, which was hauled up with several suitcases and boxes on the aforementioned train.

With my lodging I am satisfied in that it is very cheerful; my bed chamber, however, has no doors, but only quite simple curtains. With a heart full of thanks I went a few hours after my arrival to the church where one of the best preachers in the United States, namely Dr. Hamilton[2] of Mobile, preached; after the sermon, preparations were made for the Holy Communion. I looked with longing at the table on which stood the bread and the chalice, for a long time had passed without a chance being offered to me to partake of this holy sacrament. The preacher remarked that during the singing of two verses the communicants should take their designated places, which was soon done. But to my not small astonishment, the preacher now issued a very friendly invitation to all those who still wanted to participate in the Holy Communion, adding that it was not intended only for a particular religious sect but for all true Christians. This invitation to the table of the Lord was too much desired for me not to accept it, and therefore I stood up and took communion with those kind people. Today I started my investigations here, and to my joy I found that there is a profusion of the finest primeval seashells here.

Wednesday, the 22nd of January. My expectations are exceeded for I have already, during my short stay, collected about 700 of the most magnificent fossil shells. In the evening usually a large chorus of frogs is heard, and on the high, steep banks of the Alabama I work surrounded by beautiful green trees, among which the magnificent *Magnolia grandiflora* is especially distinguished, often reaching the size of our oaks. Its leaves are very similar to those of the laurel, but three to four times larger. Everywhere on the rocky slopes grow glorious fan palms and evergreens, under which bloom liverwort and violets. To make the landscape still more beautiful, a small waterfall 100 feet high tumbles from the upper part of the forest, its roar mingling during the day with the songs of various birds and toward evening with the melancholy cries of several horned owls which have their dwelling in the rock crevices; often dappled goats climb around the rocky slopes.

Friday, the 24th of January. Yesterday we had a thunderstorm and heavy rain almost all day, which pleased me very much because I could stay at the house and wash my shells, which was no small task, for every one had to be handled singly and with care. This morning the sky was quite clear, and I went to the place where I used to collect. I became so engrossed in my work that I missed lunchtime and when I had finally torn myself, I would say with force, away from my work it was already three o'clock in the afternoon.

Saturday, the 25th of January. This morning everybody here complained about the severe cold, for there was a light freeze last night, but so light that it was hardly noticeable. My shell collection has grown now to 1500 choice specimens, and still I find new species.

Friday, the 31st of January. The past night was the coldest one I have spent here; there was frost and ice ⅛ inch thick. Toward nine o'clock in the morning, however, no trace of the frost could be seen.

Saturday, the 1st of February. For some time now I had settled on today as the day to explore a cave, a distance of 5 English miles from here, to which a young man who lives here was going to lead me. For that purpose I had rented a horse from my landlord, and my friend had borrowed one from one of his acquaintances. After breakfast we started on our little trip. The weather was lovely, and I was delighted anew by the wonderful scenery and the luxuriant vegetation, especially by the beautiful fan palms which are my favorites. Our way led us through several magnificent plantations, where in the gardens narcissus and other spring flowers were in full bloom. Among other shrubs I noticed several large rosemary plants in bloom; in the woods violets exuded their fragrance, and to my surprise, I saw chestnuts which were about to bloom. We missed the right way, and so were forced to ride back 1½ English miles. We finally came to the entrance of the cave, which was in a rocky mountain at the foot of which a crystal-clear brook wound through a thicket of fan palms, various evergreen magnolias, and other trees. A great mass of long Spanish moss hung from the trees, which gave the whole a somewhat melancholy appearance. For a while I stood absorbed in this beautiful view, moved by the feeling which so often overpowered me, that I could not share all this beauty with any one of those who are so dear to me and who now probably cannot go out of the house because of cold and snow, while I enjoy spring here in its very special charm.

At the entrance of the cave my companion lighted a fire; I took the wax candles from my pocket and in a few minutes we crawled down in the eerie darkness. I soon found that there could be no thought of geological yield because the limestone which formed the rather large cave came from the same epoch as the one which overlay the shells in Claiborne. In the center of the cave flowed the brook mentioned earlier, here also crystal clear. In it, as on both sides of it, lay a mass of smaller and bigger rocks which had fallen from the ceiling of the cave. Because very many threatened to follow at any moment, there was indeed danger, so while we went of course to the end of the cave, we then returned right away to daylight.

Monday, the 3rd of February. Yesterday afternoon I proceeded to my local favorite spot, a small waterfall. I took pencil and paper with me and drew a little sketch of the place, and because I noticed that the Alabama had fallen very much, I decided on an excursion of its romantic banks, where I discovered a place with fossil oysters of such an extraordinary size that with the greatest effort I could take only six of them away. This morning it rained heavily, and because I feared that the rain would make the river rise again, I set out for the place right after breakfast. Besides the oysters, I found also a quantity of splendid rare seashells 2 inches in diameter, some of which were almost entirely undamaged. When I arrived home at noon I was drenched quite thoroughly by the rain and sweat and had to change my clothes completely. If somebody had painted me in the get-up in which I came home, it would have made a queer picture. On my back I had my heavy sack with the oysters; not only my jacket but also my vest pockets were filled with fossils; under my arm I carried the umbrella (which I could not open because I had both hands full), in one hand the stone hammer and chisel, in the other a handkerchief full of petrified starfish, and I was covered from head to foot with sand and dirt.

Wednesday, the 5th of February. Yesterday and today I was busy packing my local harvest, but I had no more peace in my room, especially since the Alabama had fallen very much again, and therefore promised new space for explorations. I was not disappointed in my expectations, for I came to a spot where I found a great number of splendid objects, some of which were quite new and of great interest. There was a slight freeze the past two nights.

Saturday, the 8th of February. Several times I had decided to finish my collecting here and go to the neighboring county, but every day I was impelled for a few hours to the steep banks of the Alabama, and daily I made new discoveries. I have now finally obtained quite a clear survey of the various local geological periods, which are defined by the layers following each other, and particularly through the organic remains contained in them. There were difficulties connected with this job, and today with much luck I overcame the last of them.

Because the Alabama is still considerably swollen, I have not had the opportunity up to now to see the stratum on which the river flows and of which Conrad,[3] who explored the local region ten years ago and who described the larger part of the local fossil shells, says the following: In the lowest stratum we find various sea conches, particularly many of those with two shells, which still have their natural connection; this seems to indicate that at the time the shells found their grave here, the upheaval of the earth was not as strong as the one which took place at the time indicated by the following layers.

I have had a chance to inspect the stratum which followed that one, and I tried to obtain many nice examples of it as well as of the one above it. This stratum consists of a white-gray soft limestone and contains a large quantity of various species of oysters, which at the bottom are very young and small and which become bigger as one climbs up the 70-foot-thick stratum. Those oysters seem to have been brought here initially by a powerful current, because the shells in the lower part of the stratum are all open and show traces of that movement. In the lower part of the aforementioned stratum are also found many other ocean shells which, however, disappear almost completely in the upper part, from which one can conclude that, at the time this layer originated, the deep ocean which the low-lying shells inhabited was transformed into a shallow sea. Here the young oysters grew and multiplied, while the other conchilia, created for a deeper ocean, did not thrive any more but gradually died out. This stratum shows itself in various spots also in the form of hard clay in which I discovered pieces of teeth of two different saurians. Here are found, as in the earlier-mentioned limestone, great masses of treelike corals such as appear only in shallow seas, which offers new evidence that during the aforementioned period a shallow sea took the place of a previous deeper one. Then suddenly the whole changed anew, which the now-following stratum indicates definitely, for, to my not small astonishment, this one consisted of an enormous mass of the same species of ocean conches which were indigenous here before the oysters and which died out while the oysters throve here for a long time, so that they grew up to be true giants of their species between forests of treelike corals. The stratum of ocean conches which lies above the oysters was 10 to 15 feet thick. The shells were not petrified but were found in their natural state; many of them still retained their color and could not have been any more beautiful if they had just come from the primeval ocean; many break, though, if one touches them. As amazing as this may appear, there exists the probability that, through a new upheaval of the earth, the shallow sea in which the aforementioned oysters and corals vegetated again was transformed into a deep ocean in whose depths the oysters found their grave and fossilized, and, through the same upheaval which caused the ruin of these oysters, a current must have been created which brought here a new colony of the descendants of the ocean conches buried under the oysters whose ancestors, just as they did, lived in a part of the earth which was spared by the revolution during which their Claiborne relatives were buried under the oysters. Approximately in the center of the stratum of these ocean conches is found a stratum only 2 to 3 feet thick in which again a few species of oysters appeared mixed with ocean conches. What greatly surprised me was that in the vegetable matter washed up from land, which at times was found in pure layers of 1 to 2 inches thick, sand and conches were noticeable in small lumps but only as far as the oysters went up, at which time the vegetable matter disappeared again without a trace. This stratum contained also sharks' teeth, which was very remarkable. The oysters did not grow there but were brought by the same current which carried the vegetable earth there, because they are all found in large shells. This highly noteworthy stratum lets one conclude that, from the beginning to the end of the comparatively short period which is indicated by it, an island or perhaps a continent with the estuary of a big river on which ebb and flow had great influence was formed. Through this current the vegetable remains, like the shells of the oysters which lived here or in a nearby bay touched by the current, were transported. As I already mentioned, however, soon all traces of it disappear, and one finds again ocean shells without the oysters which appeared here for the last time and which were survived for a long time by them. In climbing up one now becomes aware of a new large transformation. Above the sand with the splendid ocean conches lies a yellow gray, often reddish, rock layer of 2½ to 4½-foot thickness. At closer examination one finds that this rock layer consists of more or less whole or broken remains of creatures which, as if cemented together and petrified, form a rock layer resting over the shells mentioned before. The main part here is played by a genus of starfish which appears in a large mass; they have the form of a round disk with a diameter of 2 to 2½ inches; on the bottom they are quite flat, and on top, elevated about ½ inch, with magnificent designs. A sort of ananchytes also appeared in this layer, but not too often; but very abundant was a small shell which was not even as big as a millet grain, and which, observed through the magnifying glass, has a shining dark copper color. This shell was present in such abundance that it gave the rock an appearance quite its own. I have not found them here again, either in the lower or the upper layers, and it seems as if they had appeared with the starfish and died out with them. In that substance I found also several crayfish claws as well as several species of ocean shells. Above the just-described stratum lay another, 15 to 20 feet thick, which consisted of rock similar in color to the lowest limestone; only a few of the shells existing on the lower level were still to be found here, as this layer on the whole was not as rich in organic remains as the preceding one. Quite peculiar in this formation was a small creature, a little smaller than a pea, which for want of a good magnifying glass I did not dare to classify. I found it only in a small place, yet there in such a quantity that the calcareous sandstone in which it appeared seemed quite streaked. This calcareous sandstone was covered by a layer of clay which had a gray-white color and was marbled red by iron. No organic remains were found in it. This stratum was 6 to 20 feet thick and was covered over by another gravel layer which was combined with clay and iron parts; indeed I found a ½- to 1-inch-thick iron stratum between the two aforementioned layers, and the gravel had taken on quite a red color from the iron. Remarkable were the many springs found here, all of which appeared where the gravel rested on clay. The water was, as is to be expected, ferruginous but quite clear. The gravel was about 10 to 30 feet thick and covered with a thin sandy soil in which grew very luxuriant trees.

✤ *Macon, Clarksville, Coffeeville, Washington-Old-Courthouse, St. Stephens, Mobile in Alabama*

MACON, Wednesday, the 12th of February. Yesterday morning I set out on my journey to this place, which is 22 English miles from Claiborne, on foot, because a very high price was asked for a pair of horses, and besides I would have had to take somebody with me to return the horses to Claiborne, which would have cost me at least $10.00. To be sure, carrying my not exactly light traveling bag was not very pleasant, especially since I went lame on the way. However the weather was splendid, and the road led all the way through a forest now and then interrupted by cotton plantations. Among others I saw examples of the splendid *Magnolia grandiflora*, the trunks of which had a diameter of almost three feet, but the larger part of this forest consisted of magnificent pines with needles of eight to ten inches in length which, combined in tufts, decorated these grand slender trees. The magnolias and various other trees shaded the luxuriantly shining fan palms. I was highly interested in the number of trees appearing in the north of America which are found here with the southern ones; among others I noticed hornbeams and a kind of birch.

Friday, the 14th of February. Today I went on horseback, because of my lame leg, to visit an old gentleman by the name of Chapman[1] whose acquaintance I had made, and who lived four English miles from here in the direction where I intended to explore. One English mile from his house I saw the first *Zygodon* vertebra, which for three years had lain in the fireplace and served as a so-called andiron, and the people said that it withstood the destroying force of the fire better than many stones which they had used for that purpose. The man went with me into one of his hilly fields, where there were many pieces of bones which I could not investigate any closer because of my ailing foot and the nightfall. With my friendly companion I rode back to his house, for he had invited me to consider it my temporary home until I had found a place from which I could conduct my explorations. When I went to bed my foot was so swollen that long after midnight I was lying awake in great pain, but on awakening in the morning I felt much relief due to the application of some household remedies, and decided therefore to continue my journey after breakfast in the company of my friendly host. Before we mounted the horses I first had to drink a glass of honey spirit. Since it is almost perpetually spring or summer here, the bees thrive extraordinarily, and Mr. Chapman owned a large bee farm. But honey has a low price here and sales are small, so my host had hit upon his own idea of distilling part of his honey into spirits. In color this liquor is like our clear grain schnapps, but it has a more aromatic flavor.

Three English miles from Chapman's house lies Clarksville, a small isolated town; here we stopped because I had to deliver a letter of introduction to a certain Mr. Washburn,[2] in whose fields many organic remains are supposed to be found. We enjoyed a very friendly reception here. Two English miles farther on I saw a *Zygodon* vertebra which was kept between some beehives as a kind of curiosity, but the field in which it was found, and where many other organic remains were lying, was sown with oats which had already started to turn green. From here we rode now to the main habitat of the *Zygodons*. Unfortunately the owner of the estate, in front of whose house were still lying some vertebrae of which each might weigh approximately 50 pounds, had died. Of these precious remains, too, one was to be sacrificed by fire since the present temporary keeper of the plantation had put it in the fireplace. I was fortunate enough to persuade this ignorant but kindhearted man to allow me to save the one destined for the fire sacrifice from its destruction. I would have liked to start my explorations here, but I will have to wait a few weeks because the man is at present unable to accommodate me.

Macon is the county seat of Clarke County and consists only of three inns, the courthouse, the jail, some stores, and a few private houses. The whole place lies in the middle of a large pine forest. A few years ago Clarksville, seven English miles from here, was the county seat, but, since the court was moved to this place, it is completely ruined, and only two families still live there in houses which are only partly provided with windows. On the whole it is quite customary in Alabama to have no windows, even in rather well-furnished houses, but instead only shutters which are closed during the night.

Clarksville, Saturday, the 22nd of February. The large geological scene which I had anticipated so eagerly was now lying almost completely unfolded in front of my eyes, and its records I found laid open before me, a privilege which had been bestowed upon no other geognost before me. It was no wonder, then, that for the last week I have made use of every moment of this opportunity. Besides a great number of sharks' and other fish teeth, I found here also saws from a species of small sawfish which, to the best of my knowledge, are a completely new discovery; also in regard to conchylia and starfish I have already made important new discoveries here.

Sunday, the 23rd of February. The afternoon before yesterday we had heavy thunderstorms accompanied by heavy showers. During the night lightning struck close to our house with a terrible crash which shattered at one stroke three oaks standing approximately 40 to 50 feet apart from each other. The air was so full of electricity that I noticed with surprise, before and after the stroke, how a sword which belonged to my landlord and was hanging near my bed on the wall made the same movement as the pendulum of a clock, without being exposed to a draft of air or any shock.

Saturday, the 1st of March. In the first week of my stay I was able to carry the collected objects home, but now I am having trouble taking them away daily in a sack on a horse, and the prospect is that soon I will need one or more wagons to carry to safety in the evening what I have collected during the day, although I adhere to my old rule and select only the rarest. There is almost no house in the neighborhood without some sick people, but, despite all the great exertions I am daily exposed to, God has until now always preserved my health though the sun has burned me as brown as an Indian.

Tuesday, the 4th of March. Yesterday morning I rode with Colonel Washburn to Macon, where I had to stay until after twelve o'clock to wait for a young man who was going to ride with me nine more English miles to a place where, as he told me, his stepbrother about a year ago had found the larger part of a skeleton in the limestone. The way led us through truly magnificent forests. After we had left the unique pine forest which surrounds Macon we came to deciduous trees from which a soft wind carried toward us a most pleasant fragrance before we even reached them. Astonished by this, I inquired of my companion as to the source of this delightful smell and was told that it came from an evergreen creeper vine which, as I now saw, hung in 20- to 30-foot-long garlands from gorgeous magnolias, evergreens, oaks, and other splendid trees down to luxuriantly growing groups of fan palms. The dark shining green of the thick leaves, covered by the countless high yellow flowers which spread the lovely smell of the creeper vine, vied with that of the proud magnolia, the queen of the forest. Beautiful fan palms alternated with dense clusters of scarlet-red blossoming rhododendron. This grows very plentifully, is known here under the name of bear shrub, and serves the animals as a good hiding place during the short wintertime when various trees lose their leaves. Under the fan palms and rhododendron one could see at times some modest violets as well as a kind of lily similar to our forest lilies alternating with spring flowers unknown to me. The dark shining leaves of the magnolias and other evergreen trees and shrubs gave a splendid contrast to the various shades of spring green which covered the trees. Just as in the summer our gardens are beautified by red and white roses, so here a profusion of lovely crimson and various white blossoms decorated these splendid forests, which were traversed by several rushing brooks. Shortly after we arrived at the spot which I wanted to examine I found that the organic remains which had led me here had been totally ruined through the ignorance of the owner. Therefore I stayed only a very short time and was back in Macon again for dinner.

Saturday, the 8th of March. The day before yesterday I made a geognostic excursion on horseback in the company of a friend, the justice of the peace and lawyer Pickett.[3] Our road was very difficult; it went steadily through the woods over hill and dale, now through swamps and now through brooks where the horses almost had to swim, but it was not only on this day that we had to find our path in such a way but more or less so on almost all my excursions and certainly at the expense of my clothes, which were badly affected by thorns and tree branches.

Toward noon we came to a region which I had wanted to examine for a long time. We left our horses with a man who lived here with his family and initially received us somewhat coldly. We had, however, anticipated this, supplied ourselves with a bottle of brandy which soon had its effect, so that the owner of the place was so well disposed toward us that he not only had a lunch prepared for us but afterwards even offered himself as a guide. Until evening the success of our exploration was not as great as I had hoped, but suddenly the scene changed, and I had the not small pleasure of discovering a considerable part of a head with teeth, a lot of vertebrae and large pieces of other bones, all more or less embedded in limestone. At the same time shells which were completely new to me revealed themselves in this limestone. This evening I also found a very nice piece of petrified wood which has great value in that it proves my earlier assertion that once there was firm land here cut through by a river which emptied into an ocean bordering the land. We had collected such a big load that we could hardly take it to the house of our guide, more than one English mile away. I had to leave the largest part of my treasure here to have it picked up later by a wagon. Yesterday morning we continued our trip to a man who lived three English miles from our night lodging; in the course of it my otherwise good horse slipped and missed only by a hair falling with me into a deep stream. The man we had looked up was one of the best and oldest hunters of this region, and my intention was to get him to ride with us to an area which we could not have found alone and of which I expected very much. We found our man not only at home but also willing to fulfill my wishes, and because his old mother had heard me called doctor, she took me for a medical doctor and asked me to write several prescriptions in a hurry, in which I naturally could not oblige her.

Monday, the 10th of March. The day before yesterday I was on my way to visit a spot which, as I had heard, was supposed to be rich in organic remains, and reached at sundown my old friend Mr. Chapman, who invited me to spend the night, because he was going to accompany me the following morning and show me the way which, as he said, I could hardly find alone. Mr. Chapman had been my companion on many of the most difficult excursions which I had undertaken here, and many an important discovery which I made in remote wilderness I owe to him. We spent a pleasant evening talking before a bright fire in the fireplace. Yesterday after breakfast we mounted our horses and rode on a scarcely walked-on footpath through the lovely smelling forest which daily grows more glorious. Among others I noticed many strikingly beautiful bushes which were covered with fine, bright, rose-red clusters of flowers which I found to be most similar to our honeysuckle, except that these were no vines but shrubs. After we had ridden through three rapid, swollen streams and many smaller ones, and over swamp, up hill and down dale, we arrived at our destination and found that the old man whom we had looked up had told us a shameless lie, and we had therefore made our arduous trip in vain. One can easily imagine how vexed I was by this; we therefore spent only a few minutes here. On the way back we found the dorsal vertebra of a *Zygodon* and some fossil starfish, so that at least I did not have to ride home entirely empty handed. At dusk I arrived very tired in Clarksville shortly before a heavy rain.

At Tattilaba Creek,[4] the 27th of April. On the 11th of March severe rainy weather set in, lasting several days and forcing me to use the time for sorting a great number of sharks' teeth and other small fossils which I had collected in the environs of Clarksville. My intention, to be sure, was to go again, as soon as the weather changed, to the earlier-mentioned place near Tattilaba Creek to continue my investigation. Absorbed in thoughts and busy with the mentioned work, I was in the house of my friend Colonel Washburn, when a mailrider, who had to take letters between Prairie Bluff and Washington-Old-Courthouse (Alabama), arrived with a letter for the colonel. This mailrider watched me at my work and soon asked what I intended to do with the objects lying before me, whereupon I gave him a short explanation. The man listened to me with much attention and then said: if it is your intention to collect such objects here in Alabama, then you must go to Washington County, where near the local courthouse lies an enormously big petrified shark which has more bones in its skeleton than you can take away in several carts. The length of this monster, he added, I cannot estimate precisely, since only occasional parts of the bones are seen above the limestone or the earth, but its length is surely over 90 feet.

This completely unexpected information, as incredible as it seemed to me, had great importance for me, and I was at once convinced that the described skeleton could not be that of a shark because the depicted length and size of the dorsal vertebrae did not correspond to those of a shark. At the same time, this mailrider mentioned that indeed single ribs and vertebrae had been dug from the ground but had not been removed because of their weight. Traces of parts of the head he had not noticed.

This American volunteered to travel the 40 English miles with me to show me the exact spot, an offer which I accepted gladly, especially because he had told me that yesterday at the departure from Prairie Bluff he had met a peculiarly built wagon, drawn by two horses, in which sat three gentlemen and one lady who were from the State of New York 2,000 miles away. These gentlemen had remarked that they had come to this region to collect petrified shells and bones; the big skeleton in Washington County was also known to them, and they wanted to obtain it. This for me very unpleasant news moved me to ask the mailrider to inquire again in Washington-Old-Courthouse and give me an answer as soon as possible.

The mailrider returned the 13th of March to Clarksville without being able to give me satisfactory information about the fossil skeleton; however, I learned from him that the company from New York had not yet visited that place. I decided without delay to set out on the trip to Washington-Old-Courthouse,[5] and through the kindness and willingness of the colonel I received a horse which he let me have for the whole time of my search. Before my departure I particularly sought to gain the friendship of the man who had temporary supervision over the old plantation of the recently deceased Judge Creagh,[6] which, situated about six English miles from Clarksville, contains a number of valuable fossils. After I had won the man to me with some presents, I departed on March 15.[7]

The trip led me in the beginning for six to seven English miles through mountainous forests, whereupon I came to a place which a few weeks ago was visited by a terrible storm. For many long miles, but only one-half English mile wide, this storm had devastated everything. At four o'clock in the afternoon I came to the small town of Coffeeville, which numbers only about 150 inhabitants, where I had to stay because I could not have found another place before nightfall. It seems strange that this place, half an English mile from the river Tombigbee,[8] lies on a highland. Although this river is navigable for more than 100 German miles by large steamboats, and a great quantity of cotton is shipped on it annually, the valley of the river here is so unhealthy that every year many people who had settled near the river have died of yellow fever. Therefore there is in this region only a single house, close by the river, whose inhabitant keeps a ferry, plants some cotton, and carries out commissions of merchandise.

The following morning I crossed the river, rode two English miles through cotton plantations, and arrived toward noon in Washington-Old-Courthouse, which formerly was the county seat of Washington County but which has been dilapidated for some years; for the court was moved 11 English miles to the town of Pareton. The surroundings of the place are very much torn up by volcanic eruptions and submersions; the place itself lies on a not insignificant elevation. The former courthouse was a three-story house of hewn beams, approximately 45 feet long and 28 feet wide, now a ruin, but the so-called jail stands intact, not far from it. The jail cells are very dark and, like the whole building, built of beams. For me the jail had great value because through the kindness of the innkeeper it was put at my disposal during my stay in Washington County, and I could store my collected objects in a dry spot and lock them up during my absence; at the same time I was given the opportunity not only to work in the shade, but, during rainy days, in a dry place. Besides these buildings there is a three-story tavern, a store, a blacksmith shop, and a tannery.

At my arrival in Washington-Old-Courthouse I at once made inquiries about the big fossil shark, but I learned only that a little white boy and a Negro slave had talked about it, both of whom were unfortunately not to be found that day. The following morning, though, to my joy both persons presented themselves to me and offered to take me to the spot, on which, as can easily be imagined, I set foot with almost indescribable feelings.[9] The place which contained the *Hydrarchos* discovered by me lies near a not unimportant river, the Sintabouge River.[10] Surrounded by woodlands, it is an elevation of volcanic origin on which there is no wood. The larger part of the surface consisted of a yellowish limestone mass, covered in part by black-brown earth which contained, particularly in the limestone, many beautiful fossils.

According to all the geognostic explorations and research I have had a chance to conduct in the United States of North America, there existed at the time when the *Hydrarchos* and similar creatures were living in this region up to the northern states a tropical climate, and the gulfs of the then North-American coast, as well as the estuaries of the bigger rivers, were the abode of the *Hydrarchos* and innumerable sharks of different sizes. It is to be assumed that at that time here a very considerable volcanic eruption of the ocean took place and that this monster was left without water on the elevation. After the softer parts had decomposed, the bones were lying for some time unpetrified, which was proved by the fact that the spinal cord canals were filled with petrified material consisting of soft lime substance, which formerly formed the ocean bottom and now enveloped the whole thing. But before this had taken place, individual bones of the skeleton suffered displacement; indeed, a part of the ribs, the foot bones and several of the tail vertebrae were lost completely, as well as a large part of the lower jaw. Nevertheless this colossal skeleton lay there connected to the extent that it formed unmistakably a sort of half-circle. Still it was noteworthy that the head was completely turned around and that the lower jaw bone was lying approximately 1 foot from it, pressed together in an angle of 45 degrees. In the upper and lower jaw many teeth were still preserved; several had been broken out in the meantime, possibly before the petrifaction of those remains took place. The front part of the head, consisting of tender bone substance, had suffered the most, as had the upper teeth parts, but of the whole so much still existed that the missing parts could be replaced artificially. It was a peculiar circumstance that almost all dorsal vertebrae lay upright just as was their position on the animal; therefore I found these on the lower side completely undamaged, whereas I found the spinal processes more or less destroyed. The ribs were partly still quite undamaged, but I was forced to dig them out piece by piece with great caution, and in fact numbered all the pieces belonging together before I took them from their position and fastened them on specially made iron arcs.

While I was still busy extricating the remains of the *Hydrarchos* I was told that in Clark County in Mississippi in a small uncultivated prairie near the bank of the Chickasawhay,[11] there was a very large petrified skeleton. So as to leave nothing untried I made the costly and difficult journey there but found only a number of ruined dorsal vertebrae which years ago had been laid in a row by Indians, but it was not possible to determine to which creature the vertebrae once had belonged.

Earlier a geognost could have made great scientific discoveries here, for the owner of that tract of land had some years ago dug up the remains of a large creature; the bones were so big and numerous that it took several wagons to take them from the place where they were found to a lime kiln, because this man thought these remains would make quite excellent lime. In this he found himself very much deceived because the bones withstood the fire. Nevertheless the bones were lost forever for science because, when they were taken out, everything was broken in pieces.

The packing and the land transport of my fossils collected in Washington County caused me no small trouble, but this also passed, and on Tuesday, the 20th of April, in the morning I rode away again from Washington-Old-Courthouse to St. Stephens, a small place lying on the Tombigbee which several years ago was of importance, but consists only of a few houses which let one guess the former prosperity of their owners. An interesting fact connected with the history of St. Stephens is that at the time of the decline of the place, which was completed within a year, more than 50 houses were moved from here to Mobile, where they still stand today, a circumstance which probably is found very seldom or not at all in history. These houses were almost all three-story ones, built of wood only, and covered with shingles. I have been told by credible persons that the art of moving houses from one place to another has reached such perfection that almost no nail and no shingle is lost.

At noon I arrived in St. Stephens. After eating I undertook a geological examination of the over-100-foot-high rock which forms the bank of the river here and which was of particular interest to me. The heat was terribly oppressive, but I could not wait for the cool of the night. The rock consisted completely of brittle limestone lying over chalk cliffs, which occurs so frequently in Alabama, but I could not discover in it any fossils new to me; nevertheless I took pleasure in the wild romantic sight of the rock formation, which often hung so threateningly above me that I did not stay under it any longer than necessary.

From St. Stephens I intended to ride back to Clarksville. But on the bank of the Tombigbee I had again to wait a long time until I was taken across by the ferry. Here I inquired again about the road which I had to take to get to Clarksville. I rode two English miles through a cotton plantation, but then came a hitch for the road divided into three arms of which I took the one that seemed to me the right one according to the description. For three miles I had to pass through a marsh where I thought every minute the horse would sink with me; finally I reached firm ground again, but my hope of meeting someone I could ask for the right way was in vain. Not until I had ridden 14 English miles did I come to a watermill, where I met three Negroes whom I asked how far it still was to Clarksville. The black faces looked at me with astonishment, until finally one of them said I must mean Coffeeville, and so I found out that I was three English miles from that place, which I had to pass through. When I left St. Stephens in the morning I had to ride 20 English miles to reach my destination, and now that it was two o'clock in the afternoon I had ridden 14 English miles in the most oppressive heat without having eaten or drunk anything at all, and I still had 18 English miles to Clarksville before me. But hesitation was no help here, only the drive forward, and a little after seven o'clock in the evening I arrived at the home of my friend Colonel Washburn, where I was received with great joy.

The next morning I used to pack some of my things which I had brought with me from St. Stephens, and after lunch I went on my way to Tattilaba Creek in Clarke County. Here is hard life and hard work. I sleep on a straw sack and have only salt pork and bread to eat. The heat is terrible, and at my work I am completely exposed to the burning rays of the sun. There is no fodder here for my horse; it is, therefore, enclosed in a sort of pasture, and I myself have to lead it twice a day ¼ of an English mile to the water.

Monday, the 28th of April. Today it was again very hot and especially oppressive for me because I worked almost all day in a hollow where lay the head parts of the *Zygodon*, which I have now undermined as far as I dared for the present; I have also already affixed a suitable scaffolding underneath it. The many curious who come every day and who in their ignorance could break the whole thing cause me no small worry.

Wednesday, the 30th of April. My anxiety concerning the *Zygodon* head was, alas, not without reason, because yesterday while I was busy packing fossils one of these inquisitive tampered with it and broke loose a considerable part of the lower jaw. It can be imagined how great my displeasure was. Today I took the whole thing apart, since there is no other way to take it out, and because the man who had helped me for a few days was drunk. I had to carry every single piece home on my back in the most terrible heat, which was so great that some iron tools which I had used and left lying in the sun were so hot that I almost burned my fingers when I tried to pick them up.

Thursday, the 1st of May. It is really singular in what way the vertebrae of the *Zygodon* were used and destroyed here. As I earlier found several of them instead of the so-called andirons in fireplaces, so one was buried in the ground, near Clarksville, to be used as a support for a garden gate. Another one I found cemented in as a cornerstone in a chimney. Here at this place where I am just writing one serves a Negro as a pillow. So and in a similar manner were those remains of the prehistoric times snatched from science by ignorance.

Clarksville, Monday, the 5th of May. After I had left the day before yesterday John G. Creagh's plantation at Tattilaba Creek, I finally arrived here again. At some distance from Clarksville I noticed large smoke clouds rising above the tops of the tall slender pines toward the sky, so that the sun appeared like a red fireball without rays, and I began to get anxious about the kind people in that place. But soon I saw close to the road the flames run up many of the splendid pines like snakes and move quickly toward the road, which induced me to force my snorting horse to greatest speed until at a turn in the road I met a crowd of people who, to my surprise, were starting fires at the roadside, which moved crackling toward the other one. To my question whether the fire in the woods was not already large enough they answered that this was the best and surest way to keep the ominously encroaching monster from crossing the road; this, as I was soon convinced, really came true, because only a fence in the severely threatened Clarksville became a victim of the fire.

When I was still half an English mile from the place, suddenly a violent thunderstorm broke loose and let itself be heard menacingly over the tops of the tall pines, so that I spurred my horse to the greatest speed, and in two minutes reached a deserted house just as the first big drops fell. A few seconds later I would have been soaked through and through.

A mishap, many will smilingly say, which would not have been too serious, but at the present time of year it was dangerous to get wet because, only too easily, the fever could follow.

Claiborne, Thursday, the 8th of May. The day before yesterday I took my leave from the Washburn family, where I enjoyed so much friendship. I still had the colonel's horse and rode to my old friend Chapman, who was of great help and use to me during my stay in Clarke County, and who at parting showed me the great kindness of riding with me here so as to take my horse back to its home. Everything went fine, and we stayed a few minutes in the little town of Macon, where I wanted to say a few farewells. We were already 11 English miles from Macon when I noticed with dismay that I had left my manuscripts there. My old companion said quite imperturbably: well, then we have to go back, which we did at once, but for a long time I was annoyed with myself for my forgetfulness. Toward evening we arrived again in Macon, quite tired, after we had ridden about 23 English miles for nothing. Yesterday morning we left the town anew, and in the afternoon reached Claiborne safely and without special adventures. I found all my things which I had left behind in good condition.

This morning my old friend rode with two horses back to his home and I had the pleasure of visiting again my old place of discovery in Claiborne. Toward evening I went down again to the Alabama to check something in the lowest layer about which I was in doubt, but before I got there I noticed a formation which, because of the dry weather that had set in, seemed different from three months ago. I started to explore, but how great was my surprise when I discovered here several layers of the finest tree leaves, below and above which there was a thick stratum of ocean conchylia.

Tuesday, the 13th of May. Because I am now working on a report of my explorations and discoveries in Alabama, I have a chance to rest from the past great exertions, which I find very beneficial. Since my return to Claiborne the weather has been very unsettled; one day it is so hot that one could faint, the other so cold that one has to have a fire in the room, the third day it rains for a change; in the mornings the air is usually filled with unhealthy fog.

Tuesday, the 20th of May. I have finally not only packed all my natural science treasures found in Alabama and sent them to Mobile but I have also composed my report on my discoveries and research, and I can now leave Claiborne with a clear conscience.

Mobile, Thursday, the 22nd of May. Yesterday morning after ten o'clock I left Claiborne on the steamboat *Admiral.* The boat was heavily loaded with cotton and also fairly crowded with passengers. The trip on the Alabama was very interesting, but it convinced me that the banks of this river are much more beautiful in the winter than in the summer because the many creepers, trees, and bushes hide too much of the glorious rock sections. In the afternoon we had to land to take on wood, and there we heard an interesting hunting story which had happened here a day ago. The man who lived here had for nights missed several of his half-grown pigs, and because he thought that they had been stolen by a Negro, he lay in wait for the thief, accompanied by his two dogs and armed with an ax. He did not have to wait long because the thief's arrival was made known by the cries of a pig; but the farmer was not a little frightened when in the full moonlight he saw his agitated dogs pursue a big bear. The bear had dropped his screaming prey and re-treated, chased by the dogs, into the nearby Alabama. The man jumped into a boat and soon succeeded in killing the black thief with several blows on the head. Toward evening we saw an approximately seven-foot alligator which lay very quietly on the bank and did not let himself be disturbed by the steamboat.

At daybreak I went to the warehouse to see whether my collection had arrived safely in Mobile; to my relief it had. I was astonished by the filthiness of the city of Mobile. When we left the steamboat an atmosphere of horrible odors met us, permeating all the dirty streets which were bordered on both sides with green gutters, but this green did not come from plants but from decaying matter. The city was already quite empty, because whoever can leave it in the summer flees before the threatening destruction. With these prospects my decision too was quickly reached. I intend to leave here tomorrow noon and to send my collection by water to New York, while I shall continue my return trip via New Orleans.

✤ From Mobile in Alabama to St. Louis in Missouri and from there to New York

O N BOARD the Steamboat *Republic*, 90 English miles above New Orleans, Sunday, the 25th of May. As intended, I left Mobile at one o'clock the day before yesterday on the beautiful steamboat *Creole*. I was convinced that a longer stay there would have made me sick. Our boat moved fast during the day, but at night it ran aground on a sandbank from which the rising flood freed us only toward morning. In the afternoon we saw a great number of dolphins jumping in the sea, which is, as already mentioned, a sure sign of a storm coming up in 24 hours, which did in fact rise up yesterday toward evening with such force that our steamboat, although fastened at the mooring and shielded on both sides by other large steamboats, shook violently. I was glad to be on the Mississippi and not on the Gulf of Mexico any more.

Yesterday morning I landed in Louisiana, took the railroad to New Orleans, where I had my things brought right away to the steamboat *Republic*, and last evening at nine o'clock we left New Orleans for St. Louis. The steamboat *Republic* belongs neither to the fastest nor the finest boats, but it was the only one which was going to St. Louis, and so I had no other choice. In New Orleans, too, all the streets were filled with thick, evil-smelling atmosphere, which is due in large part to the filthiness prevailing there.

Monday, the 26th of May. We have just left the city of Natchez, after a stop of only 10 minutes, which time I used to run to the bank and look at some fossils which were collected here. As short as my stay was, it convinced me sufficiently that I would not have found anything new had I spent the past winter here as I had planned. For the fossils here are of the same period as the ones I collected last fall in Missouri, only with the difference that these here were washed up on shore and were very worn while the others were native and completely preserved.

Friday, the 30th of May. It is almost impossible to write on our steamboat because it makes terrible movements when in motion. The steamboats going up the Mississippi land seldom, even when firewood is loaded; the flatboats on which the sellers have stacked the wood are fastened to the steamboat while it is in full motion, and so the wood is transferred from one boat to the other. The day before yesterday we passed a spot where a great many vultures were assembled. When we came closer, I saw that they were enjoying the partly decomposed body of a drowned stag; although we passed by very close, they let themselves be disturbed but little in their feast. Yesterday we reached the estuary of the Arkansas into the Mississippi. Here lies a small city called Napoleon. The change in the climate is now quite considerable, and this morning with a clear sky it is so cold that everybody is running around rubbing his hands and trying to get warm by moving around.

Sunday, the 1st of June. Yesterday afternoon we stopped for a few minutes at the city of Cairo. A few days ago we met again a floating house with a diving bell which was anchored above the wreckage of a sunken steamboat. I heard the remark from one of our pilots that, if one counted the steamboats which were wrecked between St. Louis and Cairo in the last ten years, one would find that at at least every three English miles one had sunk. The distance is 190 English miles. In the evening we passed the very much-feared spot which bears the name "steamboat graveyard" because annually many steamboats sink here and hundreds of travelers with them; it was a dreadful thought for me in the darkness and still of the night to pass over those buried in this swirling depth.

St. Louis, in Missouri, Monday, the 2nd of June. This morning at three o'clock we finally arrived safely here. My first concern was to look for an Ohio boat, and I found that the big and beautiful *North America* would leave St. Louis this afternoon. After I had done all that was necessary for the transport of my collection to New York, and visited some old acquaintances, I boarded the ship, which departed soon afterwards.

Smithland, Wednesday, the 4th of June. Last night I arrived here very tired because I have, since I left Alabama, not slept one night in peace, for on the Mississippi one must constantly be prepared either not to wake up at all or to awaken in imminent danger to life. My intention was to go from St. Louis directly to Golconda, but the Ohio is already so low that the steamboats cannot land there. I decided therefore to let myself be taken off at Smithland. The steamboat *North*

America was a magnificent boat, not unlike a swimming palace, and it covered the 250 English miles from St. Louis to here in less than 24 hours, but it was also full of passengers, and it received another large influx at Smithland, where we met a steamboat which did not dare go up the Ohio any farther because of the low-water level.

Monday, the 9th of June. Last Thursday I was put on land one English mile from Golconda because it was impossible to land there. The next morning I rode at once to my lead mines, which as before mentioned were near here, where my friend Graham informed me what had been done here during my absence. I took samples of the various ore pieces, and after lunch we rode back to Golconda. We arrived there two hours before sundown and used the time looking for fossils, of which we indeed found some beautiful examples. The morning before yesterday Mr. Graham and I packed my collection; to strengthen the boxes we enclosed them with hoops for which we had selected suitable wood in the forest. The weather in the past days was so hot that the thermometer showed 90° Fahrenheit in the shade. Yesterday morning I made an excursion with Mr. Graham to the neighboring rocks, where we again found some splendid fossils. Here I also found again a poisonous insect which had already tormented me very much in Alabama and which I thought to have escaped now; it is the size and shape of a bedbug but much more flat and so hard that it can be killed only with difficulty. It is chocolate colored and has a yellow spot on the back. It bites the skin so circumspectly that one notices it only after the developing of a wound which in a short time swells up like a large mosquito bite.

Sunday noon, the 15th of June, on board the steamboat *Financier* above Portsmouth. Last Monday at midnight I was awakened by a Negro of my hotel in Smithland with the news that a steamboat was coming. I quickly dressed, and there had been indeed no time to lose, because I had only been on the little steamboat *Whiteville* two minutes when it continued its trip to Louisville. The captain very courteously led me to a small bedchamber furnished with two beds (an arrangement found on all American steamboats) and put at my disposal an empty bed, although the floor of the cabin was covered with sleepers and it was difficult to make one's way among them without stepping on one or the other. Enough; I had a bed and was very glad of it. Our *Whiteville* moved slowly, to be sure, but safely. Last Wednesday we met three steamboats which were stranded together on a sandbank, and on Thursday, the 12th of June, we landed safe and sound at Louisville in Kentucky. We were early enough even to go by steam ferry to Jeffersonville in Indiana, where I had left in storage the objects which I had collected last fall at the Ohio falls. I found them in good condition and had them brought to Louisville. Friday morning in Louisville I boarded the mail packet to Cincinnati, on which I should have arrived there Saturday before daybreak, but when I awoke Saturday morning I found that we were sitting quietly on a sandbank 24 miles from that place. After 24 hours' hard work the floating of the boat was achieved, and so we arrived at about ten o'clock in the morning at our destination, where I had a good opportunity to embark on the steamboat *Financier* for Pittsburg. I hardly had time to have my luggage taken from one boat to the other when the bell sounded for departure, and the wheels agitated the water as if everything was impatient to move us on.

Yesterday afternoon we saw a long procession of people moving from a small hamlet toward the Ohio. At first I could not make out what their purpose was until somebody suggested that probably they would be holding a baptism here. He was right, because soon one of the men left the people standing on the bank and went almost up to his hips into the water; he turned around again and took hold of a woman dressed in black with whom he repeated his wet promenade. When both reached the spot from which he had turned back, they stood motionless, the woman was totally submerged backwards by the man, and both then stepped back onto land. The man now fetched a second woman who was also submerged much to the delight of our boat's crew.

Wednesday, the 17th of June.[1] For two days now we had toward evening a short, severe rainstorm, and, strangely enough, both times at the moment when we were about to put passengers ashore. The day before yesterday, for example, we landed at a small place and hardly had time to secure the steamboat on shore when a storm broke loose so fast and violently that it threw a man from the boat into the water. Fortunately the Ohio was so shallow here that the victim only fell up to his armpits in the water, and that to the laughter and cheers of the onlookers, some of whom had to pay for the spectacle with their hats, which the wind tossed into the water, whereby the laughter was increased even further. Yesterday toward evening we experienced a similar scene. We had just landed to let off one of our travel companions when storm and rain broke loose with full force. We were at the mouth of a tributary of the Ohio, at the bank of which lay an ingeniously constructed ferry; it was built like a steam ferry but so small that it was set into motion by two horses, of which one on either side in a little house turned a wheel which was connected to a waterwheel. A single man managed the whole thing, for the horses were so trained that on command both started their march, and stood motionless immediately when the man rang a small bell. During the storm there appeared on the opposite bank a wagon loaded with wood and drawn by four horses. With much skill the ferry was set into motion, and despite the storm, not only reached the bank but also took the wood wagon with the four horses and delivered them safely across the river. The wagon had hardly reached land when the ferry was seized by the force of the storm and turned around. Right after the wagon had left the ferry, it had to go up a fairly steep hill, but because the horses were too weak to take their load uphill, the weight of the wagon drew them backward toward the deep river. The back wheels were already in water, and the horses, sensing their danger, summoned all their strength to keep themselves on land when the back axle broke and the wagon overturned throwing the load into the water. The driver and the team were saved, however.

New York, Saturday, the 27th of June.[2] On June the 18th we landed in Pittsburgh. To my not small surprise the city was already rising out of its ruins, even though only two months had gone by since the nicest part of it had been laid in ashes.[3] Already some larger and smaller houses had been rebuilt, and many four- and five-story ones were near completion. In the 42 streets which were devastated by fire one could pass only with difficulty between the ruins and the newly rebuilt houses. I was glad, therefore, when the canal boat on which I wanted to travel to the Allegheny mountains left in the evening. Sleeping accommodations were in short supply on this boat for there were about 60 passengers on it. But because I had immediately after my arrival in Pittsburgh taken care of my trip on the canal, I had now the advantage of belonging to those who had a choice, for it was the custom here that at bedtime the names were called in the order in which they were registered. Above me hung two beds, one above the other; during the night the one who slept right above me fell from his bed. He fortunately came to no harm,

but in falling gave me an unpleasant shock, snatched up his bed linen
and blanket, and climbed back into his bed. When it is very warm the
situation must be very oppressive, but it was our luck that the weather
was cool and everything went as well as could be expected. On the
morning of the 20th we arrived without particular adventures at the
place from which the railroad leads over the Allegheny mountains, and
within minutes we had climbed the first. We reached the mountain in
various ways, sometimes by horses, sometimes by engines, sometimes
by stationary steam engines operating cables, and finally, the coach ran
by itself 4 English miles down the mountain to the place where, for the
second time, we went aboard a canal boat. Rhododendrons in full bloom
adorned the mountains here. Plane trees had not quite developed their
leaves, and all vegetation was still in its spring dress. Two days and two
nights we traveled again on the canal and passed through many charming surroundings. The 21st of June we arrived in Harrisburg, the capital
of the State of Pennsylvania, where we again took the train. In the
afternoon we reached Philadelphia, where I stayed for two days and
made some very interesting acquaintances, among others that of Professor Dr. Morton,[4] by whom I was introduced into the local Natural
History Society. June the 24th I left Philadelphia for New York, where
traveling part of the way by steamboat and part of the way by train,
I arrived safely the same day. But a severe blow was awaiting me here.
As mentioned earlier, I was forced to send my boxes, which contained
the *Hydrarchos* as well as the rest of my Alabama collection, by ship
from Mobile to New York, although the dangers which threaten the
ships in the Gulf of Mexico were not unknown to me, but I had no
other choice. Enough; my things were loaded on to the beautiful and
almost new boat *Newark* for New York. Right after my arrival in New
York I hurried anxiously to the mercantile house to which my belongings
were addressed; one can imagine my dismay when I received the news
there that this ship had been wrecked and sunk at the Florida coast not
far from Key West. A few days passed in great anxiety, when a weak
ray of hope appeared, for in one of the New York newspapers appeared
the news that a few items had been saved by the wreckers of Key West.
Because I knew that my things were the last ones loaded, it was to be
expected that they would be among the first to be saved. But this ray
of hope also vanished with the thought that if the wreckers had indeed

saved my boxes, and found on opening them that they contained, instead of mercantile goods, only stones, they would have felt disappointed and probably would have thrown the contents back into the sea to make room for more valuable merchandise. Even the opposite situation, that the wreckers might really have an idea of the value of my objects, was not much more consoling, because I would not have been able to pay the large sum which would have been due those people under the salvage law. What a joyous surprise it was therefore for me when, a few days later, I again held in my hand a New York newspaper in which it was reported from Key West that the wreckers had held a consultation and agreed to charge the owner of the salvaged cotton 42 percent of the value as salvage reward, but to send the giant sea serpent free of all charges on the *Globe* to New York, since this fossil sea monster was a scientific object and they were well aware that the discovery of it had entailed much difficulty and expense for its finder. Conscientiously those noble-minded men kept their promise, so that some time later I had the great pleasure of exhibiting the *Hydrarchos* for the first time in that part of the world whose waters it had ruled thousands of years ago, in whose ground its remains were wondrously preserved and through whose generous people it was saved from sinking in the depths of the sea. During the exhibition in New York I was visited by the captain of the sunken ship, from whom I learned the interesting fact that at the time his ship was wrecked, the storm was raging so wildly, and the waves of the agitated sea went so high, that he had given up hope that the daring wreckers would venture on to the wreck. They had not only done this, however, but had also shown even more courage and intrepidity. For when they were loading one of my boxes into their boat, it fell overboard into the foaming sea, whereupon a wrecker, without prolonged reflection, tied one end of a long rope to the boat and, with the other end in his hand, jumped into the waves, under which he immediately disappeared. But shortly he again appeared on the ocean's surface, and was grasped by some of his comrades, while others tried to pull up the box, which was attached to the rope, from the ocean floor. Certainly this instance, in which wreckers salvaged an item which would have been lost forever without their help, not to acquire money but solely to do science a service, deserves to be generally known.[5]

Tuesday, the 4th of July.[6] Yesterday the anniversary of American independence was celebrated, and from earliest morning until latest night one heard nothing but the firing of all possible guns. At the city's expense fireworks were set off in the local park to the accompaniment of music and the cheers of a large crowd. The final and main piece represented in several variations a temple of freedom and was a masterpiece of its kind. Toward noon I went for a walk to the local so-called Battery in the harbor, where everything was still quiet except for a few half-grown boys who now and then fired overloaded pistols. Just ahead of me I noticed a three-decker with 74 cannons, which lay perfectly quiet and, apart from the American flags flying gaily in the wind from the huge masts, seemed to be quite listless. I had already turned my eyes away from it when all of a sudden it seemed to dissolve in fire while the thunder of its cannons sounded; the distant echo of this fearful thunder had hardly subsided when it was answered by a warship lying a bit farther away, and then by a third one. I had a chance here to imagine to some extent the frightfulness of a sea battle.

Thursday, the 17th of July. Yesterday I had the pleasure of greeting here again my friend and traveling companion, lawyer Ludewig from Dresden. The intense heat of this year had driven him from Washington to this place, where at least the ocean air is somewhat cool. The blazing heat is so great that many people fall dead in the streets.

Saturday, the 19th of July. This day will remain unforgettable for the citizens of New York. After fire alarms had already sounded during the night at various times, and the fires, set by incendiaries, had been extinguished before they could spread, I was again awakened this morning shortly before three o'clock by the fire alarm. But because I thought again that there was no danger, I turned over in bed with the desire to go to sleep again as soon as possible, which I was prevented from doing by the increasing ringing of all bells. I now jumped out of bed and went to the window to look at the sky reddened by the fire. All at once a gigantic column of flame surrounded by jet-black smoke rose to the sky, and at the same moment the whole big city of New York trembled with a noise as if everything was going to pieces; along with this there was such strong pressure in the air that for a moment it took my breath away. With that we had a view which could truly be called frightfully beautiful, yet I would not wish it for anybody. That this whole scene of terror had been caused by an explosion there was no doubt.

The alarm bells were silent before the explosion took place, but now they began anew with double vigor to toll through the dreadful night, and in various places the frightful devastation-threatening flames appeared, and in all the streets one heard an increasingly alarming turmoil. I had quickly dressed and was now hurrying to the fire. When I arrived, about 30 large houses were already in flames because the explosion had thrown the fire, like incendiary rockets, far and wide. Three houses had already collapsed, and almost all windows in the neighboring streets were shattered by the explosion. Because of this, the firebrands had easy access to the open windows, by which means the fire spread with such incredible speed that it is hard to imagine. In this violent wind the flames spread with such force that many families with their half-naked children in light night clothes could barely save themselves. One could see things being thrown from every window onto the streets. The fire was still quite far from the museum where I had my collection, but it worked its way with gigantic strides toward that neighborhood, and for over an hour I watched with anxiety its incredibly speedy advance; but all of a sudden the wind changed and drove the flames in the opposite direction. Only now did I dare to leave my watch and go to some other spots. The fire raged until about noon, and more than 400 big houses were laid in ashes. The damage has been estimated at over $6,000,000, and how many people may have lost their lives or become cripples is not known. Already this afternoon several families had a search made under the hot ruins for the crushed remains of their kinfolk.[7]

Wednesday, the 23rd of July. Yesterday morning I negotiated about exhibiting my *Hydrarchos* in one of the big local public places of amusement which is known by the name Niblo's Garden. The owner very amicably invited me to the evening entertainment and I had a very pleasant evening. The place comprises a not large but very tastefully laid out garden with a theater which holds about 1500 spectators and which is regularly attended. Besides this there is a fine saloon of considerable size, where all kinds of refreshments can be had. This evening a tragedy was performed first, followed by a comedy. On my way home after eleven o'clock the fire bells sounded again. At home I learned that the fire had started in the tobacco shop of two Germans who lived in the same house with me, and who had not heard about the fire, which did not do as much damage as did the water used to put it out, until the danger was already over.

Thursday, the 24th of July. This morning I went for a short while to the place of the conflagration, which, as can well be imagined, offered a very sad appearance, and where still in many places high smoke clouds rose to the sky and flames flared up, though a very considerable number of workers were busy tearing down walls and cleaning up. I particularly noticed a cellar from which leapt a bright flame, and which seemed to be in a terrific, blazing heat.

Saturday, the 26th of July. Yesterday I had the unpleasant job of having my complete Alabama collection moved from the second floor to the basement of the New York museum because a dispute had developed between the owner and the tenant, as both wanted an equal share of the profits from my exhibition. My boxes had already been thrown around so much that most of them hardly held together any more; I opened one of the boxes with uneasiness but found to my joy that the contents had suffered no harm.

Tuesday, the 7th of August.[8] On August the 2nd I walked over the burned-out places and was not a little astonished when I noticed, close to the still-smoking ruins, or rather surrounded by them, a big new building which was already built up to the fourth floor. At present I am very busy with the erection of the *Hydrarchos*, so that I have only little time left for other work.

The description of the hydrarchos which follows here is excerpted from Koch's booklet, "Description of the Hydrarchos Harlani," written for the New York exhibition of the skeleton.

This relic is without exception the largest of all fossil skeletons, found either in the old or new world. Its length being upwards of one hundred and fourteen feet, without estimating any space for the cartilage between the bones, and must, when alive, have measured over one hundred and forty feet, and its circumference probably exceeded thirty feet.

The supposition that the Hydrarchos frequently skimmed the surface of the water, with its neck and head elevated, is not only taken from the fact, that it was compelled to rise for the purpose of breathing, but more so from the great strength and size of its head, which could, with the greatest ease, be maintained in an elevated position, when in the act of carrying in its jaws a Shark or a Saurier, while struggling for life, to free itself from the dreadful grasp with which it had been elevated from its native element, to serve as a morsel to this blood thirsty monarch of the waters. The eyes of this monstrous Reptile, were situated in such a manner, as to assist it in seeking and catching its prey with more ease, for they were not only of a large size, but prominently placed on the forehead, and measured from six to eight inches in diameter. The strong and lengthy tail was used as a rudder to direct its course, as well as for the purpose of propelling. The ribs are of a very peculiar shape and form; so much so, that I know of no animal to which I might compare them; the greater number are small, and remarkably slender on their superior extremities, until we arrive within two thirds of the length toward the inferior extremities, where they begin to increase in thickness most rapidly, so that near the inferior extremities where they are flattened, they have three or four times the circumference that they have on the superior, and they have very much the curve of a sickle. From the whole of their construction, we may justly form the conclusion that the animal was not only possessed of a fleshy back of great power, but also, of remarkable strength in its belly, by which means it was enabled to perform very rapid movements notwithstanding its two paddles being quite small in comparison with the rest of the skeleton, yet they are in proportion with the short and thick Ulva and Humerus, or fore-arm, which, together with the paddles, have been concealed under the flesh, during the life of the animal, in such a manner as to be only perceptible through muscles and cartileges, similar to the fins of the eel. The Humerus and Ulva are not unlike those of the Ichthyasaurus.

The Hydrarchos has little in common with the Saurier, or Lizzard, with which a large number of monsters of old are classed, and with whose remains we have already become acquainted, through the progress of geological discoveries; as the teeth of all creatures belonging to the Saurier, or Lizzard family, have only one fang, whereas the molar teeth of the Hydrarchos have two distinct fangs; those of the anterior teeth are closely united, but become more and more forked, as we approach the posterior ones; these molar teeth have a certain analogy to those of a Marsupial animal still they are like those of all the serpent tribe; formed less for the purpose of mastication, being slight, and small in size, it would seem that the animal did not masticate its food, but gorged it entire; which is more expressly proven by the fact, that this creature was provided with palate bones, which have some similarity to molar teeth, but could only be used for the purpose of crushing its food. Its greatly elongated snout was armed with fifty or more spear shaped molars whose fangs were deeply inserted in distinct sockets. The pivocation is in the extreme anterior ones, only marked by a groove; the spear shaped crown of these teeth is divided into more or less minor spear shaped points, which increase or diminish in number, according to the situation the tooth occupies in the ramus; the central one of them is the longest, and those nearest the gum are the smallest. These crowns are covered by a thick coating of enamel, which had a rough surface, and are marked with small scale-like elevations which are narrow, lancet shaped, and elongated, with their points upward, arranged in a manner like the tiling upon a roof. All the teeth are so set in the ramus and the maxilla, that their extremities have an inclination backwards towards the palate, like those of the shark, so that the victim caught, could easily enter the mouth, but could not possibly escape. The canine teeth correspond in regard to the before-mentioned position with the molaris, as they also curve backward, as well with the superior as with their inferior extremities both of which terminate in a blunt point, the inferior being the sharpest. These teeth are from six to eight inches in length, full five sixth of their length being concealed in the ramus and maxilla; and their superior or exposed points, are covered with a thick coating of enamel, which exhibits the same marking which was observed in the already described molaris. The bodies of these teeth are compressed, and have their greatest circumference in the centre, standing from one to two inches isolated from the incisors, and from one to one and a half inches from the palate bones.

Today I saw, for me, quite a new and interesting sight. For on Broadway (main street of New York) I encountered one of the city's mail delivery coaches with 5 pairs of milk-white horses, which a single coachman guided from the box with coupling reins, and I was assured by credible persons that the same coachman had already guided 11 pairs of horses from the box all by himself.

At the shore a huge mass of people had crowded together to watch the just-arrived steamship *Great Britain*, which had made its first voyage

across the ocean. The immense ship has six masts and was decorated
with seven flags. It is made entirely of iron but on the whole has an
unpleasant appearance.

✤ Richmond, Petersburg in Virginia, Rockymount, Gaston, Enfield in North Carolina, Grove-Landing, Norfolk in Virginia, Baltimore in Maryland, Philadelphia in Pennsylvania, Albany, Utica, Trenton in New York

GASTON, WEDNESDAY, the 17th of December. On the morning of December 10th I left New York and went by train via Philadelphia to Baltimore, where I arrived in the evening, and from where, in the afternoon of the 15th, I left again in the company of a Mr. Timm to go overland to Washington. The coaches of the train which went there were truly splendid. Each held approximately 40 to 50 persons, who could make themselves comfortable in pairs on sofalike seats excellently upholstered and covered with the most beautiful scarlet cut velvet. In the center was an aisle which was closed at both ends with doors; the side walls had a row of windows which were decorated with elegant curtains. All of this was heated by a small iron stove standing in the middle.

In Washington we hired a carriage which took us and our luggage to the Potomac, a little more than an English mile away, where the steamboat *Augusta* lay ready to carry us away. After we had deposited our suitcases we went to a nearby restaurant where we had a dinner that tasted very good but which in Germany would be considered peculiar, for it consisted of oyster soup with which we drank coffee and ate pickled gherkins. I was very pleased that we had not traveled by sea, for a considerable storm had risen which soon became hurricanelike. In spite of it, we departed.

On awakening in the morning, I almost thought that our boat had set sail because it was entirely lying on its side, as I noticed in my bed; but when I went on deck I saw the two-mile-wide river so agitated by the storm that the steamboat was tossed around like a sailing ship, and I was close to getting seasick. At eight o'clock we landed at the estuary of the Potomac, which was here almost one English mile wide and deep enough for a large warship. From here we went again by train, to Fredericksburg, Virginia, in one hour. We were told that for six years there had not been such a violent storm in this region as last night. Everything was covered with snow, while in Baltimore and Washington there was none at all. It was really curious that the farther south we came, the more snow there was. When we arrived at noon in Richmond, the capital of Virginia, we learned that for the last few days communication with the farther distant South had been completely blocked because of the heavy snow that had fallen there. Already before Richmond I noticed by the various trees that we were approaching the South. Among others I saw many beautiful holly trees, whose vivid green leaves contrasted very much with the white snow, also some species of birch existing only in the South. We passed Petersburg. The snow was now so bad that the few travelers were all put into one coach in which not only the sacks containing letters and newspapers but also all the suitcases and packages had sufficient room. Only now we were in no slight danger, because to get through the snow we had to have one engine in front and one behind us. We were therefore, in the true sense of the word, between two fires, and if the slightest thing happened to one or the other engine we would be lost; however, everything went well and we arrived safely here. One of my traveling companions here told me the following anecdote. A Professor Michael, employed by the University of North Carolina, was breaking up some stones with his hammer while on a short geognostic trip to the interior, when he was arrested by the superintendent of a plantation and taken to a justice of the peace. When the professor asked what he was supposed to do before the justice of the peace, the superintendent answered that he was going to be taken to an insane asylum. Fortunately the justice was a reasonable man who invited the professor to a good lunch.

Enfield in North Carolina,[1] Sunday, the 21st of December. As a result of the exceptionally high snow, we were held up 24 hours longer than expected and did not depart from Gaston until the evening of the 17th, and that in a manner completely new to me but which was the only means of transportation available here. For we were obliged to hire a Negro slave who was supposed to transport us and our suitcases from Gaston to this place, a distance of 1 ½ English miles, on a small railroad vehicle. Our suitcases were put on one side of the small vehicle, we occupied the other, and behind us stood the Negro, who then with admirable skill pushed us ahead, first across a 500-yard-long bridge over the Roanoke, and then continually uphill to the place where the train picked us up. Without a single adventure we arrived here the day before yesterday. Enfield is a small place, consisting of a tavern, a store, and a small railroad station, completely surrounded by forest. It was the first time here in North Carolina that I had had a chance to examine the local Miocene shell marl, which is of considerable geological interest, and, for the regions where it exists, is in the economic respect of inestimable value because, on account of it, the poor sandy soil will, in time, become relatively fertile. Only since a little more than a year ago has anyone started to look here for the shell marl and use it. In regard to the variety of shells and corals occurring here, the local area, from my present observation, stands no comparison with Claiborne, in Alabama. Yesterday morning we set out for the so-called Fishing Creek, where on the property of a certain Mr. Doshyre were supposed to be the remains of a sea monster.[2] Because the road was very bad, we wanted to use the train, which went in our direction for a part of the way. The ride was spoiled for us in the following manner. It was a very cold December morning, so some black railroad workers, who rode with us on the car loaded with beams, had thrown a layer of clay on one of them and started a merry fire so as to keep warm during the ride. Waiting for the locomotive to start up I stood by the fire, when suddenly it started up with a violent jerk, and I fell full length, close to the flames. Startled, I tried to get up quickly and almost fell between the two cars; after I had safely escaped this danger, the sparks of the engine flew around our heads to such an extent that we had difficulty in keeping our clothes from being burned. We then continued on our way on foot, and found in the owner of the place we wanted to visit a very accommodating old man, but learned from him that the water level of the river was so high that we would not be able to see the slightest thing, which was a hard blow for us after such a long and costly trip.

Tuesday, the 22nd of December.[3] This morning we set out to examine a marl deposit six English miles from here which belongs to a Dr. Whitaker.[4] In doing so we experienced two new demonstrations of hospitality. After we had walked approximately one-half an English mile we encountered Mr. Whitaker, a merchant from Enfield, in a fine one-horse carriage, and greeting us cordially, he offered us the use of it while he himself was going to continue on foot to Enfield, but we declined with thanks this most courteous offer under the pretext that we preferred to walk in this cold weather. Arrived at Dr. Whitaker's, we were received very amicably, and not only did he and his adult son accompany us to his marl pit but he also had prepared in our honor a noon meal which one could have considered a banquet. The marl pit was filled to the brim with water; a great heap of marl had been thrown out, and I had the pleasure of finding in it a not small collection of fine Miocene shells. It would have been difficult for us to take our harvest to Enfield if a young doctor had not by chance seen our heavy bundles and, going the same way, kindly offered to take them with him.

Petersburg in Virginia, Sunday, the 27th of December.[5] Yesterday morning at one o'clock the passing train picked us up and we rode on it the 19 English miles to the small place, Rockymount. The railroad leading here is the worst I have traveled on in all my life. The coach often rolled like a seagoing ship, and I was afraid that at any moment it would turn over. The way was through almost continuous forest. In Rockymount a man of about fifty introduced himself as the innkeeper, but he had more the appearance of a robber-chief than that of a hotel owner, as he called himself. Discouraged by that, we had not the slightest inclination to put up at his house, but he described to us a great many advantages we would enjoy in his house and made also the only-too-true remark that, first, it would be a long way to the other hotel, and second, nobody was here who could take our heavy suitcases there. So we had no other choice than to give our things and ourselves up to this man and his black companions, one of whom lighted the way with a pine torch through the pitch-black night and the marsh. It would be too much to give a description of the house which now accommodated us, and its inhabitants, but this description would furnish material for an original piece if it were sketched by a skilled writer. Although the house was still almost new, its few windows were for the largest part deprived of their glass, which had its good side, since most of the prevailing evil smells, caused by the most terrible filthiness, were able to escape. Added to that was the fact that partly on the floor and partly in some beds lay several drunken men who called themselves gentlemen and who would have started a fight immediately if they had not been shown proper respect. We were both very tired, for it was already three o'clock in the morning, and we had not been able to sleep during the night while we were traveling, so we lay down for a few hours on a single bed assigned to both of us, which could not, despite the weak glow of our light, hide the dirt. After breakfast we went out to see whether some geological discoveries could be made, but after just a few minutes I could see by the formation existing here that we had come here in vain, and we could have turned back immediately if we had not had to wait until evening. Among the few inhabitants of Rockymount there was great excitement because five Negro slaves were going to be sold to the highest bidder, an event which had lured here many residents of the surrounding area who regarded this as a festive occasion and, naturally, more than half of those present were drunk. To get out of the way of our host and these inebriates we spent a large part of the morning and afternoon in the forest at the romantic rocky banks of the Tar, which flows, roaring loudly and at great speed, by Rockymount, and close to the road, it drives the works of a cotton mill in which only blacks are employed. In the evening the train finally appeared to rescue us from here, and so we started on our way back to Virginia.[6] The country here must be very delightful in the summer, for already in winter it is very romantic. Most of the big primitive rocks which rise in the forest in various forms are covered with vivid green moss and small fern, and I thought it very noteworthy in the middle of winter to find a pleasant variety of green trees and shrubs; among the latter the beautiful holly distinguished itself especially. They have the greatest resemblance to orangery trees not only in respect to height but also as to shape and leafiness. There are also many evergreen creepers with vivid dark leaves which climb luxuriantly on the otherwise leafless trees and bushes. The beautiful green cedars and the slender tall long-needled pines also add much to the agreeable change.

This morning we arrived safely here in Petersburg in front of Jarrat's

Hotel and at once made little excursion on the opposite bank of the Appomattox River but found that there was nothing for us to do here, for everything consisted of a formation of granite and so on, which however was partly covered with red Miocene sand but apparently containing no organic remains. However, the country was very lovely, and the city of Petersburg, which numbers 10,000 to 12,000 inhabitants, is not only very pleasantly situated but is also, on the whole, built with much taste; many houses are surrounded with charming gardens.[7] Several nice churches and towers add to the beauty of the place. We looked at the city hall, built of granite, which has a tower with a statue of the goddess of justice with the scales. Our hotel is very comfortable and contrasts strongly with the one in Rockymount.

Grove-Landing, Wednesday, the 31st of December. On the morning of the 29th we were awakened before dawn to get ready for our trip but still had to wait a long time for our breakfast, and after we had gulped it down in the greatest hurry, we jumped into a coach which took us to the railroad station at City Point [8] on the James River. At the station we were caused much annoyance because we had to wait almost an hour until two Negroes had pulled enough water from a well in a bucket to fill the locomotive, which took all the longer because almost all the water poured in ran out again. It took just as long to load the necessary firewood, which was still being sawed and chopped when we had already been sitting in the car for a long time, and when we finally left it took us 2½ hours to cover a stretch of two German miles. About 600 paces from City Point we came to a complete halt because the engine, despite all efforts, would not move, since it had to overcome a small hill. All the passengers got out now to walk the rest of the way, and everybody marched, loaded with suitcases and packages, slowly toward the river, where at any moment the arrival of the steamboat coming down the river was expected; the boat, without mercy, would have left anybody behind who was not on the banks at the time of the landing. Our suitcases were the heaviest, and we could not quite bring ourselves to this departure, but it almost looked as if there was no other choice left. Everybody looked at us, partly sympathetic, partly scornful, when the false cry "the steamboat!" sounded. We already had one of our suitcases halfway out of the car when the train slowly started to move again, probably because it did not have to pull so much of a load any more, and so we arrived with our things in due time at the landing and laughed at the others. Those who by chance had with them black slaves, heavily loaded with baggage, urged them on to hurry with cajolery and threats, while the ones who lacked such black servants dragged, panting along in various groups, with suitcases, travel sacks, and so on. The whole thing afforded a very comical sight until the steamboat *Alica* arrived from Richmond, took us aboard, and after a trip of 45 English miles put us ashore at Grove-Landing.

My idea was, when we left Petersburg, to land at the little town of Jamestown, which lies on an island in the James River and was founded by Captain Smith, who discovered the river in the year 1607 and who is well known for having been caught by the Indians, and while being led to his death was saved by the beautiful Indian maiden Pocahontas.[9] Jamestown is now completely deserted, and only the ruins of a small church are still to be seen. The captain of the *Alica* told me that if I wanted to get off here I would have no other transportation; therefore I went 10 English miles farther until I arrived with my friend at the so-called Grove-Landing. I was one of the first to jump on land, so as to get a wagon immediately which could take us and our heavy suitcases to the town of Williamsburg, 7 English miles from here, where, according to Mr. Lyell, we had to continue our local geological investigations. I had hardly set my foot on the bank and cast an inquiring glance around when I forgot all about the wagon and began to look for a house where we could stay at least a few days, for I immediately realized that this was the place we had been looking for. Already here the stretch of sand bordering the wide river was strewn with partly broken, partly whole fossil seashells which belonged to the Miocene formation and which had been washed out in the course of time from the 50- to 60-foot-high river banks. The latter consisted almost entirely of stratified shells which, joined together by sandy marl, formed a firm mass covered with sandy clay on which grew coniferous and deciduous trees. From time to time pieces of the high banks fell on the sandbanks where then, of course, trees, shells, and pieces of marl lie topsy-turvy but are soon washed away by the waves, with the exception of most of the trees, which cover the low sandbanks in all directions and make walking very difficult, indeed frequently almost impossible. The river has a pronounced ebb and flow, and when the latter occurs its lower banks are quite inaccessible.

To transport the newly arrived travelers, a number of wagons drawn by horses and mules were waiting; these belonged, like the black, woolly-haired drivers, to the owner of the local plantation, Mr. Vinu, who sat in a small cabin and wrote down the names of those who wanted to avail themselves of his wagons, and at the same time accepted payment therefor. When I asked for shelter, I too was referred to this gentleman, and I put to him briefly my request and the purpose of my visit, where-upon, without ceremony, he invited me to his house and promised to have our suitcases taken there, which truly took a great load off my mind. Now that I had quarters, I started at once with my research. It was already dark when we arrived at the house of our present host, which was situated a little down the river on a terracelike elevation and afforded a delightful view across the river, which here is seven English miles wide, to the opposite bank. Often one could see a number of ships, large and small, which went up and down the river, which provided a special charm to the whole scene. The house itself was almost 100 years old; it had been built by a rich Englishman, entirely in old English style, and it will no doubt stand for another few hundred years, which in America is truly something extraordinary. The stones of which it was built had all been brought here from England, and all the woodwork was done in old English style and very painstakingly.

Norfolk, the 6th of January 1846. I have during the past days, New Year's day not excepted, worked very hard and strenuously but achieved my aim and gathered together a magnificent collection. Yesterday, together with Mr. Timm, I was busy packing it; for that purpose we had earlier collected a quantity of soft, short, dead grass, because no cotton was to be had here as it had been in Alabama, and ordinary hay was too coarse for our purpose. I also tried in vain to buy boxes in Williamsburg, but fortunately I obtained some through my host. About 20 minutes after we had packed everything, a steamboat came down the river and landed for a few minutes at Grove-Landing to take us and some other passengers of different colors on board. We had a very pleasant trip and arrived at sundown in Norfolk. This town numbers approximately 15,000 inhabitants, and across from it lies the smaller town of Portsmouth, in which the fine St. Mary's Hospital looks espe-cially beautiful. The building and interior cost $900,000; the United States also have here a very large shipyard. I did not like Norfolk very much; to be sure, the most unpleasant rainy weather we encountered here contributed much to that. There was a period when Norfolk had more ships than New York, but that golden time is long past, and almost all buildings which are near the pier look old and dilapidated. The main street, though, has rather pretty houses; on it is also the courthouse, which has much similarity to an old German city hall in a small city. Because today there was a court hearing in the murder of a sailor two days ago, we entered the courtroom, whose interior corresponded with the exterior. After we had waited a long time, the court finally started with the questioning of the witnesses. The criminal was a young man of about twenty-four years, with a bloated face which promised little good. His left eye was completely bloodshot, and the area around it was almost entirely black, probably from a terrible blow which he had received from his victim; incidentally, not the slightest traces of remorse were to be detected in him. When, after some time, we went from the courtroom into the street, we heard music which was far from sad, and saw passing the funeral procession of the murdered man, whom many of his comrades, a band in the lead, were escorting to his place of rest.

Baltimore, the 11th of January. On the 7th we left Norfolk on a steamboat. Because it soon got dark and it still rained heavily, I could see little of the far distant banks of the river and, since a violent storm was approaching, I went to bed. I had not yet fallen asleep when our boat dropped anchor because our captain did not dare to go on. Toward ten o'clock in the evening the storm abated a little, and the sky seemed to clear up, which induced our captain to weigh anchor and move forward again, but we had scarcely left our anchoring for half an hour when the storm arose with renewed force. I had fallen asleep during this time and started to dream that I was sailing on the ocean in a terrible storm which was so violent that I, being seasick, had the greatest difficulty staying in my bed; indeed, many of the crashes were so violent that I was thrown against the ceiling. Finally I awoke and found that my dream was, to a large extent, reality. To be sure, I was not on the open sea, and also not on a sailing ship, but I was actually seasick, and that on a steamboat which in this jet-black night during the most frightful storm battled with the furious waves, while the cranking of all beams and the howling of the wind made horrible music. Finally toward morning the storm abated again, and at the same time we entered the narrow part of the bay. On January the 7th at noon we reached, without further mishap, Baltimore, from where I sent my travel companion Mr. Timm back to New York.

Yesterday toward evening I took a walk to the local telegraph office, which stands at the edge of town on a 50-foot-high hill and allows a wide view over the entrance of the harbor. With surprise I noticed here a line of business which some local Germans have been conducting fairly successfully for several years. The hill on which the telegraph office stands is formed of various kinds of soil, of which the lowest one is a coarse white sand which is covered by several layers of a bluish ferruginous clay. The sand is of such quality that it can be mixed with slaked lime for use in building. These Germans have turned this to advantage, and they have undermined the hill so that it has a great similarity to a colossal rabbit furrow which is traversed by 8-foot-high and 3½-to-4-foot-wide passages which go a considerable distance into the hill and branch into several arms.

Philadelphia, Sunday, the 18th of January. On Monday the 12th of January I disassembled my *Hydrarchos,* on exhibition [in Baltimore] until now, and soon had it packed, but because steamboating in the winter is very irregular, I had to wait until the 14th of January before I could transfer it to a steamboat leaving for Philadelphia. These steamboats are also better equipped for cargo than for passengers; but, because I wanted to stay with my boxes, I left Baltimore with one of them in the afternoon. These steamboats run from Baltimore a distance of 60 miles along the Baltimore Bay, and from there into a canal which connects it with the Delaware Bay. On Thursday at sunrise I landed in Philadelphia, very glad to leave this boat. My first business here was to have the boxes with the *Hydrarchos* brought to the local museum, the largest and most beautiful in the United States,[10] and yesterday the assembling was completed.

Utica in New York, the 7th of April. Early the 27th of March I left Philadelphia to travel to New York, where I arrived safely the same day. It was conspicuous to observe that the vegetation in New York was considerably behind that of Philadelphia; in New York I also still saw ice in several streets, which was no longer to be found in Philadelphia. On the evening of April the 3rd I left New York by steamboat and arrived safely the next morning in Albany. There I made the acquaintance of a Mr. Meyer, the local agent of a society of German emigrants, which numbers approximately 700 persons and which had settled six English miles from the city of Buffalo; 600 of them are already here, and the rest will soon follow from Hesse, which they had to leave because of their religious beliefs. They call themselves "Inspired Ones" and their main purpose is to come as close to the first followers or disciples of our Savior in every respect as possible. They own 7,000 acres of good land, and each of them has equal ownership. These people seem to like their new situation, although they had much to contend with, and, as can be expected, will have to overcome many an obstacle still. The land on which they now live they bought from the Indians, because it was a so-called Indian reservation.

Yesterday noon I reached Utica. Here too was still much snow, and half the street in front of my hotel was still covered with ice 1 to 2½ feet thick. Utica, for its size, is the most beautiful place in the State of New York; it has approximately 15,000 inhabitants, nice wide streets and very tastefully built homes, almost all of which give evidence of the prosperity of the inhabitants. The canal which leads from Albany to Buffalo cuts across the center of town.

The 9th of April. The day before yesterday I seated myself in the mail coach, which I was going to take up to the town of Trenton. I was told that the road was bad almost beyond comprehension, and I found it to be only too true. For that reason the post did not consist of a coach but of a long heavy wagon covered with oil cloth and drawn by four horses. The road was a series of deep and dangerous marshy holes covered with round boulders, large and small, which appeared frequently in such numbers that, when the wagon drove between them and the marshy ground, one expected to turn over at any moment, or that the wagon would break. On both sides the deep ditches were still filled with snow, which had been so deep this past winter that one could see the ruins of four buildings crushed by the weight of it. At seven o'clock I finally arrived exhausted in Trenton. I was still eating dinner when I heard the ringing of the church bells and learned that this was the invitation to a temperance club meeting where a Virginian was going to give a lecture. Curiosity impelled me too into the meeting hall, which was lighted by twelve tallow candles, and before a quarter of an hour had passed almost all the seats were filled with men and women of various ages. The speaker exaggereated, as is often the case on such occasions; among other things, he said that he believed the final judgment on Judas Iscariot would turn out to be milder than on the brandy merchants. At the end of his edifying speech (which lasted approximately an hour) he remarked that he had gold medals with him, which, however, as it turned out later, were only gilded, of which he would present one to any drunkard who wanted to reform; but those wanting to buy such a medal could have one for one shilling, because through the sale of these he defrayed his traveling expenses. But he did a poor business, and, further, nobody came forward as a repentant drunkard.

The next morning I went 2½ English miles to the Trenton falls, which had drawn me here. They consist of several waterfalls of the West Canada Creek, and are famous not only in a geological respect for the trilobites occurring here but also because of the romantic situation; they are visited all summer by numerous tourists, many of whom stay for weeks. Cut through the lovely forest of fir and deciduous trees are several ways which lead to the most beautiful views of the waterfalls, as well as to the stairs leading to the foot of the falls. The largest of these falls is 45 feet high and approximately 500 to 600 feet wide, and bordered on both sides by black limestone which is covered with big and small firs, ferns, moss, wintergreen, and so on. They are partly enveloped in a thick mist by the rising water-spray of the wildly frothing and thundering fall. The water rolls rapidly between the high black rock walls, and then, plunging anew into the depths, changes into foam and spray, a spectacle which repeats itself four times in a short distance. Right at the slope of the first and biggest of the falls is built an elegant house, from which a wooden staircase leads to the foot of the fall. It is intended for the enjoyment of tourists in the summer. Because of the melted snow, the river unfortunately was still so high that I found the rock layers which contained the type of fossils I was looking for covered with water. Yet unexpectedly I found a spot which seems to have remained unnoticed and which provided me with a fairly rich yield of fine petrifactions, especially many trilobite heads. With this treasure I boarded, at eleven o'clock in the evening at Trenton Village, the mail wagon back to Utica, where we arrived exhausted toward one o'clock in the morning after a body-shaking ride.

The 11th of April. Still feeling pain in all joints and very tired, I went unusually early to bed on the evening of the 9th; then after midnight the frightfully horrible dirgelike sounds of fire bells awakened me. Climbing out of bed and stepping to the window, I saw that the fire must be, if not in the museum, close by it. I was not a little alarmed by it because in the next few days I was going to erect my *Hydrarchos* there, and so a serious disruption, as well as a not small loss, was in store for me. With these thoughts I dressed and went to the scene of the fire. There I found to my relief that the fire was some houses away. A considerable crowd had, as usual on these occasions, gathered, and also some pumps were there, but to my not small surprise the whole gathering seemed to consist only of curious spectators, since even the usually very active firemen stood calmly by in their uniforms looking at the fire which just now caught the shingled roof of the second house. Only the people occupied with clearing out were busy; it even looked as if they would abandon the house undisturbed to the flames. The matter was almost inexplicable to me, until I finally learned that there was no water for extinguishing the fire because the water from the nearby canal had been drained for repair work. Finally, the house in which the fire started collapsed, and at the same time one pump received a little water from a well, whereby the burning roof of the next house was wetted until more water came, and so this house could be saved.

Friday, the 17th of April. For several days I have been busy examining the slate deposits existing close to town here, which in the geological survey of the State of New York were specified as Utica slate and which are supposed to be very poor in fossils. Yet I discovered that there are four times as many fossil species in it as were found and described by the state geologist.

Sunday, the 19th of April. Last evening I went with the owner of the local museum, who was awaiting the arrival of the *Hydrarchos* as anxiously as I was, to the local telegraph office to ask whether my things from Philadelphia had arrived in Albany, a distance of 93 English miles from here, and had been sent from there to this place. The question was written on a piece of paper and handed to the manager of the telegraph office, who at once, without any preparation, took it in his left hand and while reading it, touched with the index finger of his right hand, at great speed, a small peg approximately two inches in length and the thickness of a thin pencil with a flat button on top. This took about half a minute, whereupon he turned to us and said: "Your question has arrived in Albany," and the last of these few words still hovered on his lips when a paper roll attached to a wheel started to move and delivered in a kind of cipher the answer: "I have completely understood, and shall send somebody to the agent of the canal boats." This first answer followed approximately one-eighth of a minute after the question. I almost could not believe my eyes, for I had expected that it would take at least several minutes until an answer could come back. So four to five minutes had passed when again came the answer from Albany. According to the canal boat agent, who had been sent for during this time, my things had not yet arrived there. This news was of course as distasteful to me as it was unexpected, but I knew now where I stood and could take measures accordingly. The telegraph between Albany and New York was not yet finished, or I could have had news in a few minutes from Philadelphia, because between those two principal cities such a wonder communication already existed.[11]

Tuesday, the 21st of April. Because my *Hydrarchos* still had not arrived yesterday morning I decided on a geological excursion on which Mr. Timm, my business assistant, accompanied me. A local geologist had recommended a quarry four English miles from here, called Black Stones Quarry, very rich in petrifactions, and toward this place we directed our steps. As chance would have it, we were directed another way from the one we should have taken, and so we soon left the road to reach our destination over hill and dale and through fields and woods. After we had climbed the first mountain, I found some exceptionally nice petrifactions, and the thought of making an unusually rich find in Black Stones Quarry began to come alive in me. It was almost noon when we finally, after long rambling and climbing about, arrived at our actual destination; but a few minutes were enough to convince me that almost nothing was to be found here for me. We therefore turned back soon.

Sunday, the 26th of April. Breaks in the canals are causing much confusion in mercantile traffic again this spring. Because of this, for example, the canal between Philadelphia and New York had not been opened until the 16th of April, instead of on April 1st. Since the 24th of this month the local canal has been navigable again, but only to about 10 English miles below Rochester, where a few days ago another serious new break occurred, which will take probably ten days to repair. Three hundred canal boats had accumulated below the break between here and Albany, and it can indeed be imagined what an interruption this mishap caused and still causes in local business.

My *Hydrarchos* had arrived in Albany on April 23, but could not be loaded because there was no empty canal boat. This totally unexpected delay was extremely disagreeable to me, but nothing could be done about it.

This afternoon I took a walk to the state mental hospital 2½ English miles from here, a splendid building which was built and is supported by the State of New York. It is four stories high and has about 2,000 windows. One front is built of sandstone and decorated with beautiful columns, but the two long wings, as well as the opposite front, consist of brick. Up in the center of the main front is a glass dome from which one must have a very lovely view over the delightful countryside.

Sunday, the 2nd of May.[12] This morning on a walk in a rather distant part of the city I saw at the canal a gathering of people in whose midst I noticed a speaker standing on a pile of lumber. To hear, if possible, something of his speech, I went close, and the first words I heard convinced me that he was a so-called temperance advocate; he tried hard to convince his listeners that on the 18th of this month they must vote for the proposition that in the State of New York alcoholic beverages could no longer be sold by the portion. This day had been designated to have this question decided by a majority vote.

On the 4th of May I finally received my so eagerly awaited *Hydrarchos* in Utica, which I left on May the 28th to embark from New York, with my complete fossil collection, for my fatherland. Arrived in New York, I found that on June the 5th or 6th the Hamburg packet ship *Miles* would be sailing for Hamburg, which was very agreeable, since I had enough time left to prepare for the journey. On the evening of June 5 I went on board ship, which weighed anchor the next morning at ten o'clock and headed, with a favorable wind, into the open sea.

✤ Return Trip to Europe

IN ALL we were only two cabin passengers and nobody in steerage, which was completely filled with freight. Our captain's name was Vogler, and I found him to be a very pleasant companion as well as a dexterous and obliging man. The first mate was an acquaintance of mine, since he had done duty as second mate on the Hamburg ship *Howard* on which I had traveled to America two years before. We had a relatively long trip until we came to the banks of Newfoundland. After arrival we found cold thick fog which prevails almost always in this region, and judging by the exceptionally cold air, we were for 24 hours very close to the floating icebergs; the ship's watch was therefore doubled, but we had a fairly fast and favorable wind which in a short time took us out of range of the dangerous ice. I was seasick only briefly on this trip, and a quite unexpected pleasure was afforded me by the fact that I could read uninterruptedly for many hours on the ship without noticing the slightest discomfort. I was further favored since the captain had a beautiful small library on board which he graciously put at my disposal. Shortly after we had passed Newfoundland I sat one evening outside the cabin reading when the captain said to me: "This is a perfect evening to expect waterspouts." But because waterspouts are a very uncommon phenomenon I considered the captain's remark as just casually dropped and did not let myself be disturbed in my reading. Only a few minutes had passed when the roaring voice of the captain gave the order to the crew to lower all sails, and when I asked him, astonished, what had happened, he pointed in a southwesterly direction to the sky, where for the first time I saw the frightfully beautiful spectacle of an approaching waterspout. The air was very thick and heavy, and a narrow cloud, still hovering in the distance, extended like a giant ruler over the horizon, and from it down to the surface of the sea reached a black spectrelike figure quite in the shape of an hourglass. At the same time the stillness of the sea was interrupted by a roar sounding like muffled thunder. Everybody on the ship was now in quiet suspense, and the scene became more and more frightening, for the horrible waterspout was coming toward us in a straight line, and already in the distance we saw how the ocean water was thrown so high that it met the cloudburst descending from above. Only somebody who has seen such a frightful vision approach, and who could neither get away from it nor do the slightest thing to ward it off, can imagine my feelings when the captain assured me that in this case no human help was possible, since we had no cannons with which to shoot at the waterspout, which was drawing nearer and nearer. Then we received help from Him who had so often saved me from danger where human power had failed, for all at once the natural phenomenon which a moment ago had threatened to swallow us disappeared, and in less than half an hour a brisk wind, completely favorable to us, came up. It not only stayed with us until it had brought us to the channel separating England from France but it also let us cover, in the time of 36 hours, the 90 German miles of water so very dangerous to the seafarer. Joyous and in good spirits, we now sailed through the familiar North Sea enjoying the warm summer air and bright sky which prevailed here. This lasted until the evening of the 5th of July, when I went to bed toward ten o'clock with the blissful feeling that we were only 10 German miles from Heligoland and therefore, before the next nightfall, we would be anchored safely on the friendly banks of the Elbe. The evening was nice and clear, and the sails were moved only by a light summer wind. Carefree, I soon fell asleep with the most agreeable feelings, when, on the stroke of twelve, I was awakened by a tremendous motion of the ship, which was accompanied by a crash, as if all its joints were coming apart. Alarmed, I was about to climb out of bed when I heard the first mate, who rushed pale as death into the cabin, call in a heartrending tone of utter despair for the captain. It can perhaps be imagined how I felt, for I expected nothing else than, in a few minutes, to sink forever with my fellow sufferers to the depth of the ocean, but, as great as our peril was during this time of terror, for the moment we were saved. Completely unexpectedly, a terrible whirlwind had struck us and had seized us with such irresistible force that, had not all three masts with almost all the rigging been broken and torn away at one blow, the ship would have been upside down in the true sense of the word. Because the ship now had no more support, it was thrown back and forth, and the whole deck was awash with seawater. For about an hour I stood clinging firmly to the railing of the staircase which led to a small upper deck where, shielded from the waves which were breaking over the ship, I could watch the work of the sailors who were busy throwing the remains of the masts, the torn ropes and sails and so forth, overboard. But this view was so sad for me, and my position so tiring, that I climbed back into bed, not to sleep, but to thank God for our rescue and to ask Him for protection in the future. Five anxious hours had passed so when the captain and the mates came to the cabin. Only now did I dare ask the captain what our fate would be, whereupon he answered: "I have now done everything humanly possible. In our condition we must without resistance go where the wind or the current carries us; if the wind drives us directly into the Elbe, we are saved, provided we do not run aground before I am able to erect a jury mast." He who has traveled on the sea can best judge how little consolation these words were, but I did not want to bother the already deeply troubled captain with any more questions. Hardly 15 minutes had passed when the drooping little weather vane, made of feathers and fastened to a pole, began to rise, pointing exactly in the direction which was the only one promising safety. Everybody wondered whether the wonderfully gratifying promise of the little vane would be fulfilled, and, while the sailors tried hard to set the pieces of sail we still had on the stumps of the broken masts as well as possible, just behind us the wind rose ever stronger, so that to our delight we began to move forward quite splendidly. Encouraged by this, the sailors worked with doubled strength improving our few sails. Then a sailor, who sat on top of the broken mainmast, cried that he could see Heligoland in the foggy distance, and because the wind stayed favorable we could see the island clearly from the deck before an hour had passed. The waves were now as high as a house, but always behind us, and before the noon hour we passed the first so-called light ship, and soon we had a pilot on board who toward evening brought us safely to Cuxhaven, where we were congratulated from all sides on our remarkable rescue. We had not been long on the safe Elbe when our saving wind turned into a violent storm.

During the night we anchored, and on July 7 at two o'clock in the afternoon, thanking God, I stepped on land in Hamburg, where I stayed only until the evening of July 8. I used this time to deliver the *Hydrar-*

chos and my other things to a shipping agent for transport to Dresden. I myself went on the Magdeburg steamboat to Magdeburg, from where I proceeded without the slightest delay to Dresden and hurried into the arms of my family.

Sources Consulted
Notes
Index

Sources Consulted

Adams, James Truslow. *The March of Democracy*. New York, and London: Charles Scribner's Sons, 1932.

Alabama Secretary of State's Commission Register. Vol. 2, 1832–44. Vol. 3, 1844–67.

Alabama State Gazetteer and Business Directory, 1887–1888. Vol. 3. Atlanta: R. L. Polk & Company, 1888.

American Philosophical Society Proceedings. Vol. 1, no. 13, Vol. 2, no. 19, 1841.

Amory, Cleveland. *The Proper Bostonians*. New York: E. P. Dutton & Company, 1947.

Andrews, Wayne, ed. *Concise Dictionary of American History*. New York: Charles Scribner's Sons, 1962.

Arndt, Karl J. R. and May E. Olson. *Deutsch-Amerikanische Zeitungen und Zeitschriften 1732–1955, Geschichte und Bibliographie*. Heidelberg: Quelle & Meyer Verlag, 1961.

Barber, John Warner. *Historical Collections, Being a General Collection of Interesting Facts, Traditions, Biographical Sketches, Anecdotes, &c., Relating to the History and Antiquities of Every Town in Massachusetts, with Geographical Descriptions*. Worcester: Warren Lazell, 1844. (Given as *Historical Collections of Massachusetts* in the Notes and legend copy.)

Bell, U. R., ed. *Kentucky, A Guide to the Bluegrass State*. American Guide Series. New York: Hastings House, 1947.

Beyer, M. and L. Koch. *Amerikanische Reisen*. 4 vols. Leipzig: Immanuel Mueller, 1839–41.

Billington, Ray Allen, ed. *Massachusetts, A Guide to Its Places and People*. American Guide Series. Boston: Houghton Mifflin Company, 1937.

Bitterfelder Allgemeiner Anzeiger. 1927.

Bjorkman, Edwin, ed. *North Carolina, A Guide to the Old North State*. American Guide Series. Chapel Hill: University of North Carolina Press, 1939.

Bogle, Victor M. "New Albany, 'A Flourishing Place,'" *Indiana Magazine of History* 49, no. 1, March 1953, pp. 1–15.

———."A Society Develops in New Albany," *Indiana Magazine of History* 49, no. 2, June 1953, pp. 173–190.

Sources Consulted

Booth, Mary L. *History of the City of New York.* New York, E. P. Dutton & Company, 1880.

Boyer, Mary Joan. *Jefferson County, Missouri, in Story and Pictures.* Imperial, Mo.: Mary Joan Boyer, 1958.

Bridgwater, William and Elizabeth J. Sherwood, eds. *The Columbia Encyclopedia.* 2d ed. New York: Columbia University Press, 1950.

Carrington, Richard. *Mermaids and Mastodons: A Book of Natural and Unnatural History.* New York: Rinehart & Company, Inc., 1957.

Chapman, Carl H. and Eleanor F. *Indians and Archaeology of Missouri.* Missouri Handbook Number 6. Columbia: University of Missouri Press, 1964.

Churchman, M. S. B. *Families of Steele and Bright.* Privately issued, 1957.

Complete Guide to the St. Louis Museum, Including a Description of the Great Antediluvian Monster, the Zeuglodon. St. Louis: R. P. Studley and Company (printers), 1859.

Cooke, Donald E. *Atlas of the Presidents.* Revised edition. Maplewood, N.J.: Hammond Inc., 1967.

The Concise Dictionary of National Biography. 2 vols. London: Oxford University Press, 1965.

Corkan, Lloyd A. M. "The Beaver and Lake Erie Canal," *The Western Pennsylvania Historical Magazine* 17, September 1934, pp. 175–88.

Dana, James D. *The Geological Story Briefly Told.* New York, Cincinnati, Chicago: American Book Company, 1895.

———. *Manual of Geology: Treating of the Principles of the Science with Special Reference to American Geological History.* Rev. ed. New York: Ivison, Blakeman, Taylor & Company, 1871.

———. "On Dr. Koch's Evidence with Regard to the Contemporaneity of Man and the Mastodon in Missouri," *American Journal of Science and Arts* 9, May 1875, pp. 335–46.

———. "Zeuglodon Cetoides" and "Mastodon Giganteus," *American Journal of Science* 2, no. 4, July 1846, pp. 129–33.

Darwin, Charles. *The Descent of Man and Selection in Relation to Sex.* Second edition, revised and augmented. New York and London: D. Appleton and Company, 1922.

Derby, John B., ed. *Connecticut, A Guide to Its Roads, Lore, and People.* American Guide Series. Boston: Houghton Mifflin Company, 1938.

Der Kleine Brockhaus. 2 vols. Wiesbaden: Eberhard Brockhaus, 1951.

Dickens, Charles. *Martin Chuzzlewit.* Puck Edition. New York: Exchange Printing Company, 1868.

Edwards, Richard and Menra Hopewell. *Edwards's Great West and Her Commercial Metropolis, Embracing a General View of the West, and a Com-*

plete History of St. Louis, from the Landing of Ligueste, in 1764, to the Present Time; with Portraits and Biographies of Some of the Old Settlers, and Many of the Most Prominent Business Men. St. Louis: Edwards's Monthly, 1860.

Encyclopedia Britannica. 14th ed., 24 vols. Chicago: Encyclopedia Britannica, 1939.

The Farmers' Cabinet, and American Herd-Book 6, August 1841–July 1842. Philadelphia: Kimber & Sharpless, 1842.

Foster, J. W. *Pre-Historic Races of the United States of America.* Chicago: S. C. Griggs & Company; London: Trubner & Company, 1873.

Gidley, James W. "A Recently Mounted Zeuglodon Skeleton in the United States National Museum," *Proceedings U.S. National Museum* 44, no. 1975, 1913, pp. 649–54.

Golconda (Illinois) *Herald-Enterprise.* July 11, 1907.

Griffith, Katharine and Will. *Spotlight on Egypt.* Carbondale, Ill.: Egypt Book House, 1946.

Gross, Hugo. "Mastodons, Mammoths, and Man in America," *Bulletin of Texas Archaeological and Paleontological Society* 22, October 1951, pp. 101–31.

Harlan, Dr. Richard. "Description of the Bones of a New Fossil Animal of the Order Edentata," *American Journal of Science and Arts* 44, no. 1, 1842, pp. 69–80.

Harris, Paul S. *Fourteen Seasons of Art Accessions in Kentucky 1947 to 1960.* Louisville, Ky: J. B. Speed Art Museum, 1960.

Harrison, James L., comp. *Biographical Directory of the American Congress 1774–1949.* Washington: Government Printing Office, 1950.

History of the Ohio Falls Cities. 2 vols. Cleveland: L. A. Williams & Company, 1882.

Hitchens, Harold L., ed. *Illinois, A Descriptive and Historical Guide.* American Guide Series. Revised with additions. Chicago: A. C. McClurg & Company, 1947.

Hodge, J. W., ed. *The United States Biographical Dictionary and Portrait Gallery of Eminent and Self-Made Men.* Missouri volume. New York, Chicago, St. Louis, and Kansas City: United States Biographical Publishing Company, 1878.

Hopewell, M. *Report of the Fourth Annual Fair of the St. Louis Agricultural & Mechanical Association.* September 1959. St. Louis: George Knapp & Company, 1860.

Hopkins, Joseph G. E., ed. *Concise Dictionary of American Biography.* New York: Charles Scribner's Sons, 1964.

Howe, Henry. *Historical Collections of Ohio in Two Volumes.* Published by the

146

Sources Consulted

State of Ohio. Cincinnati: C. J. Krehbiel & Company, Printers and Binders, 1904.

————. *Historical Collections of Virginia; Containing a Collection of the Most Interesting Facts, Traditions, Biographical Sketches, Anecdotes, &c., Relating to Its History and Antiquities, Together with Geographical and Statistical Descriptions.* Charleston, S.C. W. R. Babcock, 1847.

Hunter, Louis C. *Steamboats on the Western Rivers.* Cambridge, Mass.: Harvard University Press, 1949.

Illustrated London News. November 4, 1848.

Ireland, Norma Olin. *Index to Scientists of the World from Ancient to Modern Times: Biographies and Portraits.* Boston: F. W. Faxon Company, Inc., 1962.

Jackson, Donald, ed. *Black Hawk: An Autobiography.* Second printing. Urbana: University of Illinois Press, 1956.

Jenkins, Warren. *The Ohio Gazetteer and Traveler's Guide.* Columbus: Isaac N. Whiting, 1837.

Kargau, E. D. *St. Louis in Frueheren Jahren.* St. Louis: E. D. Kargau (publisher), Aug. Wiebusch & Sohn Printing Company, 1893.

Kellogg, Remington. *A Review of the Archaeoceti.* Publication number 482. Washington: Carnegie Institution of Washington, 1936.

King, Moses, ed. *King's Handbook of New York City.* 2d ed. Boston: Moses King, 1893.

Klein, Benjamin and Eleanor. *The Ohio River Handbook and Picture Album.* Cincinnati: Young and Klein Inc., 1950.

Koch, Albert C. "Das Skelet des Zeuglodon macrospondylus," *Haidinger's Naturwissenschaftliche Abhandlungen* 4, pt. 1, 1851, pp. 53–64.

————. *Description of the Family of Animals now Extinct, but Known to the Scientific World Under the Appellation of Hydrachen; these animals, when living, were the most gigantic, powerful and horrible beasts of prey that ever ruled over and spread terror through the primitive oceans; also, an account of the discovery of the remains of Hydrachen in general, and particularly of the Zeuglodon Macrospondylus, of Mueller.* 12 pages. New Orleans: printed at the office of the *Daily True Delta,* 1853.

————. *Description of the Hydrarchos Harlani: (Koch): (The name Sillimanii, is changed to Harlani, by the particular desire of Professor Silliman.) A Gigantic Fossil Reptile: lately discovered by the author, in the State of Alabama, March, 1845. Together with some geological observations made on different formations of the rocks, during a geological tour through the eastern, western, and southern parts of the United States, in the years 1844–1845.* Published by Dr. Albert C. Koch, corresponding member of the societies of Halle, Dresden, &c., 2nd ed., 24 pages. New York: B. Owen, printer, 1845.

————. "Description of the Missouri Leviathan, together with its supposed habits, and Indian Traditions concerning the location from whence it was exhumed," *The Farmers' Cabinet, and American Herd-Book* 6 August 1841–July 1842. Philadelphia: Kimber & Sharpless, 1842.

————. *Description of the Missourium, or Missouri Leviathan; together with its supposed habits, Indian traditions concerning the location from whence it was exhumed; also, comparisons of the Whale, Crocodile, and Missourium with the Leviathan, as described in the 41st chapter of the Book of Job.* 16 pages. St. Louis: Charles Keemle, printer, 1841.

————. *Description of the Missourium, or Missouri Leviathan; together with its supposed habits, Indian traditions concerning the location from whence it was exhumed; also, comparisons of the Whale, Crocodile, and Missourium with the Leviathan, as described in the 41st chapter of the Book of Job.* 2nd ed., enl. 20 pages. Louisville, K: Prentice & Weissinger, printers, 1841.

————. *Description of the Missourium, or Missouri Leviathan; together with its supposed habits, and Indian traditions concerning the location from whence it was exhumed; also comparisons of the Whale, Crocodile, and Missourium with the Leviathan, as described in 41st chapter of the Book of Job, and a Catalogue of the whole fossil collection.* 3rd ed., enl., 24 pages. London: E. Fisher, 1841.

————. *Description of the Missourium Theristocaulodon (Koch), or Missouri Leviathan (Leviathan Missouriensis), together with its supposed habits and Indian traditions; also, comparisons of the Whale, Crocodile, and Missourium with the Leviathan, as described in the 41st chapter of the Book of Job.* 5th ed., enl., 28 pages. Dublin: C. Crookes, printer, 1843.

————. *Die Riesenthiere der Urwelt oder das neuentdeckte Missourium Theristocaulodon (Sichelzahn aus Missouri) und die Mastodontoiden im Allgemeinen und Besondern, nebst Beweisen, dass viele, uns durch ihre Ueberreste bekannt gewordene Thiere nicht praeadamitisch, sondern Zeitgenossen des Menschengeschlechts waren.* 99 pages, 8 plates. Berlin: Verlag von Alexander Duncker, 1845.

————. *Die sechs Schoepfungstage oder die Mosaische Schoepfungsgeschichte in vollem Einklange mit der Geognosie, nebst einer kurz gefassten Naturgeschichte der merkwuerdigsten Geschoepfe der Urwelt.* 51 pages. Vienna: Mechitharisten-Buchdruckerei, 1852.

————. *Hydrargos, or Great Sea Serpent of Alabama, 114 feet in length, 7,500 lbs. weight, now exhibiting at the Apollo Saloon, 410 Broadway. Admit-*

Sources Consulted

---

tance 25 cents.—Description of the Hydrargos Sillimanii (Koch). A Gigantic Fossil Reptile, or Sea Serpent: lately discovered by the author in the State of Alabama, March, 1845. Together with some geological observations made on different formations of the rocks, during a geological tour through the eastern, western, and southern parts of the United States, in the years 1844–1845. 16 pages. Published by Dr. Albert C. Koch, corresponding member of the societies of Halle, and of Dresden, &c., New York, 1845.

———. *Kurze Bemerkungen ueber die aus mehreren Arten bestehende Familie der Hydrarchen, der groessten und gewaltigsten Raubthiere der Urwelt. Nebst einigen Worten ueber die Auffindung des grossen, zu jener Familie gehoerenden Zeuglodon, welches 1848 vom Verfasser in Alabama gefunden und von da zuerst nach Dresden gebracht wurde. Mit einer zweiten Abtheilung, enthaltend einige in moeglichster Kuerze erzaehlte Kampfscenen der Indianer mit den weissen Ansiedlern Amerika's.* 32 pages. Dresden: Druck der Koenigliche Hofbuchdruckerei von C. C. Meinhold und Soehne, 1848.

———. *Kurze Beschreibung des Hydrarchos Harlani (Koch) eines riesenmaessigen Meerungeheuers und dessen Entdeckung in Alabama in Nordamerika im Fruehjahr 1845. Nebst einigen geognostischen Bemerkungen verschiedener Felsengebilde, welche der Verfasser waehrend seiner 2 ¼ Jahr langen wissenschaftlichen Reise in den westlichen und suedlichen Theilen der Vereinigten Staaten untersuchte.* 20 pages. Dresden: Hofbuchdruckerei von C. C. Meinhold & Soehnen, 1846.

———. "Mastodon Remains, in the State of Missouri, together with Evidences of the Existence of Man contemparaneously [sic] with the Mastodon," *Transactions of the Academy of Science of St. Louis* 1, 1857, pp. 61–64.

———. "On the genus Tetracaulodon," *Proceedings of the Geological Society of London* 3, pt. 2, 1842, pp. 714–16.

———. "Remains of the Mastodon in Missouri," *American Journal of Science* 37, 1839, pp. 191–92.

———. *A Short Description of Fossil Remains, found in the State of Missouri by the author.* 8 pages, 1 plate. St. Louis: Albert C. Koch, publisher; Churchill & Stewart, printers, 1840.

Kresensky, Raymond, ed. *Iowa, A Guide to the Hawkeye State.* American Guide Series. New York: Viking Press, 1945.

Kurtz, Seymour, ed. *The New York Times Encyclopedic Almanac 1970.* New York: The New York Times Company, 1969.

Lesley, C. C., ed. *Pennsylvania, A Guide to the Keystone State.* American Guide Series. New York: Oxford University Press, 1940.

Lewis, Henry. *The Valley of the Mississippi Illustrated.* Translated by A. Hermina Poatgieter, edited by Bertha L. Heilbron, St. Paul: Minnesota Historical Society, 1967.

Lloyd, James T. *Lloyd's Steamboat Directory, and Disasters on the Western Waters.* Cincinnati: James T. Lloyd & Company, 1856.

Lucas, Frederic A. *Animals of the Past.* New York: McClure, Phillips & Company, 1901.

Lyell, Sir Charles. *A Second Visit to the United States of North America.* 2 vols. New York: Harper and Brothers, 1850.

McCracken, Harold. *George Catlin and the Old Frontier.* New York: Bonanza Books, 1959.

McDermott, John Francis. "Dr. Koch's Wonderful Fossils," *Missouri Historical Society Bulletin* 4, no. 4, July 1948, pp. 233–56.

————. "Museums in Early St. Louis," *Missouri Historical Society Bulletin* 4, no. 3, April 1948, pp. 129–38.

————. "William Clark's Museum Once More," *Missouri Historical Society Bulletin* 16, no. 2, January 1960, pp. 130–33.

Mantell, Gideon Algernon. *The Medals of Creation; or First Lessons in Geology, and the Study of Organic Remains.* 2 vols. London: Henry G. Bohn [1853].

Mehl, M. G. *Missouri's Ice Age Animals.* Rolla, Mo.: State of Missouri, Division of Geological Survey and Water Resources, 1962.

Merrill, George P. *The First One Hundred Years of American Geology.* New Haven: Yale University Press; London: Humphrey Milford and Oxford University Press, 1924.

Missouri Argus (St. Louis). August 30, November 19, December 6, 1838.

Missouri Republican (St. Louis). April 18, 20, 21, 22, 25, 26, 1837; August 27, June 22, 1839.

Missouri Saturday News (St. Louis). February 10, August 11, 1838.

Montagu, M. F. Ashley and C. Bernard Peterson. "The Earliest Account of the Association of Human Artifacts with Fossil Mammals in North America," *Proceedings of the American Philosophical Society* 87, no. 5, May 1944, pp. 407–19.

Mooney, James. *Handbook of American Indians North of Mexico, in Two Parts.* Bulletin number 30, Smithsonian Institution, Bureau of American Ethnology. Washington: Government Printing Office, 1910.

Morison, Samuel Eliot. *The Oxford History of the American People.* New York: Oxford University Press, 1965.

Munro, Robert. *Archaeology and False Antiquities.* Philadelphia: George W. Jacobs & Company, n.d.

National Cyclopaedia of American Biography. 51 vols. New York: James T. White & Company, 1892.

New York Shipping and Commercial List. 1845.

New-York Commercial Advertiser. July 1, 1845.

Nickles, John M. "Geologic Literature on North America 1785–1918," *United States Geological Survey.* Bulletin 746. Washington: Government Printing Office, 1923, pp. 614–15.

Onions, C. T., ed. *The Oxford Universal Dictionary.* 3rd ed. rev. with addenda. Oxford: Clarendon Press, 1955.

Osborn, Henry Fairfield. *The Age of Mammals in Europe, Asia and North America.* New York: The Macmillan Company, 1910.

———. *The Proboscidea.* 2 vols. New York: American Museum of Natural History, 1936.

Oudemans, A. C. *The Great Sea-Serpent, A Historical and Critical Treatise. With the reports of 187 Appearances (including those of the Appendix), the Suppositions and Suggestions of Scientific and Non-Scientific Persons, and the Author's Conclusions.* Published by the author. Leiden: E. J. Brill; London: Luzac & Company, 1892.

Owen, Richard. *A History of British Fossil Mammals, and Birds.* London: John van Voorst, 1846.

———. "Report on the Missourium now exhibiting at the Egyptian Hall, with an inquiry into the claims of the Tetracaulodon to generic distinction," *Proceedings of the Geological Society of London* 3, pt. 2, no. 87, 1842, pp. 689–95.

Palmer, Katherine V. W. "Tales of Ancient Whales," *Nature Magazine,* April 1942, pp. 213–14, 221.

Richardson, Eudora Ramsay, ed. *Virginia, A Guide to the Old Dominion.* American Guide Series. Fourth printing. New York: Oxford University Press, 1947.

Richman, Irving B. "Keokuk," *Dictionary of American Biography.* Edited by Dumas Malone. Vol. 10. New York: Charles Scribner's Sons, 1933.

St. Louis Anzeiger des Westens. June 29, 1839.

St. Louis Daily Commercial Bulletin. September 1–23, 1836; March 31, 1840.

St. Louis Daily Pennant. April 16, 1840.

The St. Louis Directory for the Year 1836–37. Also for 1838–39, 1840–41, 1842, 1845, 1847, 1848. St. Louis: Charles Keemle.

The St. Louis Directory for the Years 1854–55. St. Louis: Chambers & Knapp.

St. Louis Directory for the Year 1857. Also for 1859. St. Louis: Robert V. Kennedy.

St. Louis Directory for 1864. Also for 1865. St. Louis: Richard Edwards, editor and publisher.

St. Louis Directory for 1866. Also for 1867. St. Louis: Edwards, Freenough & Deved.

Sanford, Laura G. *The History of Erie County, Pennsylvania, from the First Settlement.* Revised and enlarged edition. Erie, Pa: J. B. Lippincott & Company, 1894.

Scharf, J. Thomas. *History of St. Louis City and County.* 2 vols. Philadelphia: Louis H. Everts & Company, 1883.

Scharf, J. Thomas and Thompson Westcott. *History of Philadelphia 1609–1884.* 3 vols. Philadelphia: Louis H. Everts & Company, 1884.

Schnake, Friedrich. "Geschichte der deutschen Bevoelkerung und der deutschen Presse von St. Louis und Umgegend," *Der Deutsche Pionier* 3, no. 9, November 1871, pp. 272–77; no. 10, December 1871, pp. 299–305.

Scott, William Berryman. "American Elephant Myths," *Scribner's Magazine* 1, April 1887, pp. 469–78.

Silverberg, Robert. *Scientists and Scoundrels: A Book of Hoaxes.* New York: Thomas Y. Crowell Company, 1965.

Smith, George Washington. *History of Southern Illinois.* Chicago and New York: Lewis Publishing Company, 1912.

Stoler, Mildred C. "Manuscripts in Indiana State Library," *Indiana Magazine of History* 27, no. 3, September 1931, pp. 236–39.

Tatum, Howell. "First Alabama River Survey 1814," *Alabama Historical Quarterly* 19, no. 2, Summer 1957, pp. 218–19.

Tebbel, John and Keith Jennison. *The American Indian Wars.* New York: Harper & Brothers, 1960.

Transactions of the Academy of Science of St. Louis. Vol. 1, 1856–1860. Vol. 2, 1861–1868.

Notes

Introduction

1. Albert C. Koch, *Reise durch einen Theil der Vereinigten Staaten von Nordamerika*, pp. 107–8 (hereafter cited as *Reise*).
2. A measure of land in Old English times and later; primarily, the amount required by one free family and its dependents; defined as being as much land as could be tilled with one plough in a year. *The Oxford Universal Dictionary.*
3. Emil Obst, *Bitterfelder Allgemeiner Anzeiger*, Beilage 230 der *Heimischen Scholle*, 1927, reproduced in a letter from Rat des Kreises Bitterfeld, Bezirk Halle, Deutsche Demokratische Republik, to Ernst A. Stadler, September 3, 1969.
4. Obituary of Dr. Albert C. Koch, *Missouri Republican*, January 14, 1868.
5. *St. Louis Anzeiger des Westens*, June 29, 1839.
6. Koch, *Reise*, p. 40.
7. Obituary of Dr. James A. Koch, *Golconda* (Illinois) *Herald-Enterprise*, July 11, 1907.
8. *St. Louis Daily Commercial Bulletin*, daily except Sunday, September 1 through 23, 1836.
9. Obituary of Dr. James A. Koch; *St. Louis Directory*, 1838–39.
10. *St. Louis Daily Commercial Bulletin*, September 1–23, 1836; *Missouri Republican*, April 18, 20, 21, 22, 25, 26, 1837.
11. *Missouri Argus*, August 30, November 19, December 6, 1838; *Missouri Republican*, August 27, 1838, June 22, 1839; *St. Louis Daily Commercial Bulletin*, March 31, 1840; *St. Louis Daily Pennant*, April 16, 1840.
12. Koch, *Die Riesenthiere der Urwelt*, p. 69.
13. Ibid., pp. 69–74.
14. Listed in *Catalogue of the Fossil Mammalia in the British Museum* (N.H.), 4, 1886, p. 345; reproduced in a letter from Mr. M. J. Rowlands, Librarian of the British Museum (Natural History), to Ernst A. Stadler, March 18, 1970.
15. *The Farmers' Cabinet, and American Herd-Book* 6, 1841–42, p. 146.
16. Ibid., pp. 121–24, 128, 146, 175–76.
17. Paul Beck Goddard, 1811–66, physician, anatomist, and pioneer in photography. *The Concise Dictionary of American Biography*, p. 347.
18. Dr. Richard Harlan, 1796–1843, a naturalist and physician whose major interests were zoology and vertebrate paleontology. His most notable work was *Fauna Americana* (1825), the first systematic treatise on American mammals. *The Concise Dictionary of American Biography*, p. 400.

19. Dr. Richard Harlan, "Description of the Bones of a New Fossil Animal of the Order Edentata," *American Journal of Science and Arts* 44, no. 1, 1842, pp. 69–80.

20. Richard Owen, "Report on the Missourium now exhibiting at the Egyptian Hall, with an inquiry into the claims of the Tetracaulodon to generic distinction," *Proceedings of the Geological Society of London* 3, pt. 2, no. 87, 1842, pp. 689–95.

21. Robert Edmond Grant, 1793–1874, professor of comparative anatomy and zoology at the University of London, 1827–74, Swiney lecturer on geology at the British Museum, friend of Charles Darwin. (*The Concise Dictionary of National Biography*, pt. 1, p. 524; Koch, *Die Riesenthiere der Urwelt*, p. 24.)

22. Koch, *Description of the Missourium Theristocaulodon.*

23. Letter from M. J. Rowlands to Ernst A. Stadler, March 18, 1970.

24. Hugo Gross, "Mastodons, Mammoths, and Man in America," *Bulletin of Texas Archaeological and Paleontological Society* 22, October 1951, pp. 105, 124.

25. Gideon Algernon Mantell, 1790–1852, author of *The Wonders of Geology*, 1838, and other works setting forth his extensive investigations and discoveries. (*The Concise Dictionary of National Biography*, pt. 1, p. 837.)

26. *Illustrated London News*, November 4, 1848.

27. Obituary of Dr. James A. Koch.

28. The diary referred to is Koch's *Reise durch einen Theil der Vereinigten Staaten*, translated in this volume as *A Journey Through a Part of the United States.*

29. Koch, *Description of the Hydrarchos Harlani.*

30. A more complete set of remains of *Basilosaurus*, consisting of a part of a head, portions of the limbs, ribs and backbone about 65 feet long, was shipped from Alabama to Albany, New York, in 1842. Scientists have found that this yoked-toothed whale reached a length of 55 to 65 feet. Katherine V. W. Palmer, "Tales of Ancient Whales," *Nature Magazine*, April 1942, pp. 214, 221.

31. Carl Gustav Carus, 1789–1869, director of the Medical-Surgical Academy of Dresden, who was especially interested in comparative anatomy and craniology. *Der Kleine Brockhaus* 1, p. 190.

32. Johannes Peter Mueller, 1801–1858, Berlin physiologist and anatomist. Bridgwater, *The Columbia Encyclopedia*, p. 1339.

33. Koch, *Reise*, p. 160.

34. Letter from Dr. K. Fischer, Museum fuer Naturkunde an der Humboldt-Universitaet zu Berlin, Deutsche Demokratische Republik, to Ernst A. Stadler, May 6, 1970.

35. Remington Kellogg, *A Review of the Archaeoceti*, p. 5.

36. Koch, *Description of the Family of Animals now Extinct*, p. 7.

37. Ibid., back cover.

38. Remington Kellogg, *A Review of the Archaeoceti*, p. 5.

39. *Transactions of the Academy of Science*, April 21, 1856, p. 16; May 19, 1856, p. 17; June 2, 1856, pp. 17, 18.

40. Ibid., January 26, 1857, pp. 34–35; February 9, 1857, p. 37; Koch, "Mastodon Remains, in the State of Missouri, together with Evidences of the Existence of Man contemparaneously [*sic*] with the Mastodon," n.d., pp. 61–64.

41. Robert Widmar, husband of Koch's daughter Maria, established the *Mississippi-*

Handelszeitung as publisher and editor in 1857; associate editors were Koch and Joseph Bauer. The only German newspaper west of New York which could be called a thorough commercial journal, it was a weekly paper with a large circulation. After its destruction by fire in 1861 it became the *St. Louis Journal of Commerce.* Will of Albert C. Koch, October 8, 1864, St. Louis Probate Court, Number 8416; Edwards and Hopewell, *Edwards's Great West*, p. 166; Karl J. R. Arndt and May E. Olson, *Deutsch-Amerikanische Zeitungen und Zeitschriften 1732–1955*, p. 264.

42. M. Hopewell, *Report of the Fourth Annual Fair*, p. 148.

43. J. W. Hodge, *The United States Biographical Dictionary*, Missouri volume, pp. 402–3.

44. Letter from The Honorable R. Gerald Trampe, Associate Judge, Circuit Court of Illinois at Golconda, to Ernst A. Stadler, November 4, 1969.

45. James D. Dana, *Manual of Geology*, p. 652; idem, *The Geological Story*, p. 267.

46. Charles Darwin, *The Descent of Man*, p. 620.

47. Jacques Boucher de Crevecoeur de Perthes, 1788–1868, French antiquarian largely responsible for establishing the existence of prehistoric man. While not the first to find evidence of a Stone-Age culture, Boucher de Perthes was the first to grasp its revolutionary significance, in an age when Archbishop Ussher's date of 4004 B.C. was widely accepted as the time of creation. In 1838 he presented tools in evidence before the scientific society at Abbeville but met with disbelief, and his monograph, *De l'industrie primitive* (1846), was ignored. *Encyclopedia Britannica* (1970), vol. 4, p. 9.

48. John Wells Foster, *Pre-Historic Races of the United States of America*, p. 62.

49. M. F. Ashley Montagu and C. Bernard Peterson, "The Earliest Account of the Association of Human Artifacts with Fossil Mammals in North America," *Proceedings of the American Philosophical Society* 87, no. 5, May 1944, pp. 409–10.

50. Carl H. and Eleanor F. Chapman, *Indians and Archaeology of Missouri*, p. 25.

51. Remington Kellogg, *A Review of the Archaeoceti*, p. 5.

52. Letter from Dr. K. Fischer, Berlin, to Ernst A. Stadler, May 6, 1970.

53. In correspondence with Ernst A. Stadler in September and October 1969, William Hoffmann of the Hoffmann Funeral Chapel at Golconda, Illinois, recounts this story as it was told to him by his father, who witnessed the removal of the remains to the new cemetery.

New Haven, Hartford, Springfield, Boston, New Bedford

1. Grove Street Cemetery, the first burial ground in the United States to be laid out in family lots (John B. Derby, *Connecticut, A Guide to Its Roads, Lore, and People*, p. 237).

2. Yale College, founded in 1701 as Collegiate School at Killingworth, now Clinton; located in 1702 at Saybrook, and in 1716 transferred to New Haven. The official title "Yale University" was not adopted until 1887. (*Ibid.*, p. 239.)

3. Benjamin Silliman, 1779–1864, chemist, geologist, and naturalist; professor of chemistry and natural history at Yale. He founded *the American Journal of Science and Arts.* (Joseph G. E. Hopkins, *Concise Dictionary of American Biography*, p. 958.)

156

Notes

4. Mrs. Harrison Gray Otis, Jr., called the "reigning queen of Boston Society" (Cleveland Amory, *The Proper Bostonians*, pp. 262, 348).
5. On June 17, 1825, the cornerstone of an obelisk was laid on the battleground by General Lafayette, to commemorate the battle fought fifty years earlier (John Warner Barber, *Historical Collections of Massachusetts*, p. 373).
6. Tisbury, principally the village of Vineyard Haven, was called Holmes Hole until 1871. (*Encyclopedia Britannica*, 14th ed., Vol. 14, p. 982.)

Martha's Vineyard

1. Koch says in a later entry (Thursday, August 1) that these were Piequatto (Pequot) Indians. The Pequot (contraction of Paquatauog, "destroyers") were an Algonquian tribe of Connecticut, and one of the most feared and powerful tribes in southern New England. In 1637, the Pequots became involved in a war with the English settlers, who had secured the help or neutrality of the neighboring tribes before they marched against the Pequots. The English set fire to the Pequots' main stockade fort in a dawn raid; estimates of the Pequot men, women, and children killed varied from 600 to 1,000, while the English lost only two men. The Pequots ceased to exist as a nation after this bloodletting. Many became slaves to the colonists, while others were sold as slaves in the West Indies. Some scattered and settled with other neighboring tribes, but they were forbidden to call themselves Pequots. (Smithsonian Institution, Bureau of American Ethnology, Bulletin 30, *Handbook of American Indians North of Mexico*, pt. 2, pp. 229–30; John Tebbel and Keith Jennison, *The American Indian Wars*, pp. 22–29.)
2. Spelled Lowman's-land in the original German edition. This is a tiny island lying eight miles offshore. (Ray Allen Billington, *Massachusetts, A Guide to Its Places and People*, p. 558.)
3. According to John W. Barber, "Gay Head is inhabited by descendants of the native Indians who own there 2,400 acres of land, most of which is under good improvement. Their dwelling-houses, upwards of 35, are mostly one story, and are comfortably built. The number of their population is 235. Their church, which at present is of the Baptist denomination, is 148 years old, since the organization, and now consists of 47 communicants. Their present minister is Rev. Joseph Amos, an Indian, of Marshpee, entirely blind, but a preacher of considerable ingenuity. Within a few years the condition of these people has much improved in point of temperance and general moral reformation" (John Warner Barber, *Historical Collections of Massachusetts*, 150).
4. Spelled Wihigan in the original German edition. Since Koch had spent several years in Michigan, it seems likely that Michigan, rather than Wihigan, is what he wrote in his journal; presumably the German typesetter, unfamiliar with American place names, transliterated this one inaccurately from Koch's Gothic script.
5. Sir Charles Lyell, 1797–1875, professor of geology, King's College, London; author of *Travels in North America, with Geological Observations (1845)* and *A Second Visit to the United States of North America* (1850). (*The Concise Dictionary of National Biography*, pt. 1, p. 802.)

6. George Nixon Briggs, 1796–1861, fifteenth Governor of Massachusetts; he served from 1843 to 1851 and was reelected seven times (*National Cyclopaedia of American Biography*, Vol. 1, p. 114).

Holmes Hole, Boston

1. Spelled Whiman in the original German edition. Jeffries Wyman, M.D., 1814–74, considered the leading American anatomist of his time (Joseph G. E. Hopkins, *Concise Dictionary of American Biography*, pp. 1261–62).

New Haven, New York

1. Spelled Sheppart in the original German edition. Charles Upham Shepard, 1804–86, mineralogist, lecturer in science at Yale, South Carolina Medical College, and Amherst (Joseph G. E. Hopkins, *Concise Dictionary of American Biography*, p. 948).

From Troy in Pennsylvania to Cincinnati in Ohio

1. Here Koch meant Troy, New York.
2. Koch's description of this event is vague, and no account of it has been found in available newspapers within a few weeks of this entry.
3. The Reverend Robert Reid, uncle of Koch's wife Elizabeth, was born in Ireland November 5, 1781, and emigrated with his father and brothers to America in 1798. Ordained in 1812 as pastor of the Associate Reformed Presbyterian Church in Erie, Pennsylvania, he often officiated as chaplain to Perry's fleet in Erie and to the army on shore during the War of 1812. It was his custom to visit every family of his congregation once in every six weeks. He died on May 16, 1844. (Laura G. Sanford, *The History of Erie County, Pennsylvania, from the First Settlement*, pp. 202–4.)
4. For a complete account of the routing and building of this waterway, see Lloyd A. M. Corkan, "The Beaver and Lake Erie Canal," *Western Pennsylvania Historical Magazine* 17, September 1934, pp. 175–88.
5. Saxon Switzerland is the name for a picturesque region with steep and craggy sandstone formations extending from Pirna, Saxony, in eastern Germany, to the Czechoslovakian border.
6. Spelled Leicken in the original German edition. The Licking River rises in Kentucky and flows into the Ohio opposite Cincinnati. An important means of travel for Indians and pioneers and later a busy trade route, the river has been the setting for more than one historical happening; at its mouth in 1780 gathered George Rogers Clark's frontiersmen for their march up the Little Miami, and the battle of Blue Licks (1782) occurred in the Licking Valley. (Bridgwater, *The Columbia Encyclopedia*, p. 1131.)
7. Circleville, Pickaway County (Ohio) seat, laid out in 1810. The town is on the site of ancient fortifications of the Mound Builders, one of which, having been circular, accounts for the name of the place. The old courthouse, built in the form of an octagon

and destroyed in 1841, stood in the center of the circle. Over the years, the citizens abandoned the unusual old street plan and squared the circle. Caleb Atwater, in his *Archaeologia Americana*, described the forts at length. (Henry Howe, *Historical Collections of Ohio*, Vol. 2, pp. 411–16.)

8. Caleb Atwater, 1778–1867, pioneer and author. He settled in Ohio in 1815, practiced law and was elected to the Ohio legislature in 1821. He supported the construction of highways and canals and the provision of popular education and was one of three commissioners appointed by President Andrew Jackson to treat with the Indians at Prairie du Chien in 1829. In 1838 he published *A History of the State of Ohio, Natural and Civil.* A social and intellectual pioneer of the Middle West, and perhaps the first advocate of forest conservation, he was the first historian of his state and the founder of its school system. (Joseph G. E. Hopkins, *Concise Dictionary of American Biography*, p. 33.)

Louisville in Kentucky; Jeffersonville in Indiana

1. Spelled Klapp in the original German edition. Dr. Asahel Clapp, the first resident physician of New Albany, married a daughter of one of the proprietors of the town. He established a drug business and was active as a geologist and botanist. He kept a diary from 1819 to 1862 in which he recorded the weather and accounts of his geological trips. (Mildred C. Stoler, "Manuscripts in Indiana State Library," *Indiana Magazine of History* 27, p. 238; Victor M. Bogle, "New Albany, 'A Flourishing Place,'" ibid. 49, p. 4, and "A Society Develops in New Albany, ibid. 49, p. 176.) The Clapp diary in the Indiana State Library includes these entries: "Sun. Sept. 15, 1844—Fair. Went to the falls and visited Mr. Koch at Jeffersonville. Friday Sept. 20—Visited Mr. Koch at Jeffersonville. Thur. Oct. 3—Went to Louisville and Jeffersonville. Synod of Presb. Church commenced at this place today. Sat. Oct. 5—Explored the Falls with Mr. Koch. Thurs. Oct. 10—Explored the Falls with Mr. Koch. Tues. Oct. 29—Fair. Went up the falls and visited Mr. Koch." (Entries supplied by Mrs. Frances B. Macdonald, Manuscript Librarian, Indiana State Library, Indianapolis, Indiana, in a letter to Ernst A. Stadler, June 19, 1970.)

2. This was the Indiana State Prison South, situated on one of the outlots of the extinct town of Clarksville just beyond the line of Jeffersonville. The convicts were leased to contractors, who paid a per diem for each man employed, while the discipline, control, and personal care of the men was in the hands of a warden and other officers representing the state. (*History of the Ohio Falls Cities*, Vol. 2, pp. 464–66.)

3. The 1844 election was influenced by the "Manifest Destiny" doctrine, which made for a lively campaign. The annexation of the Oregon Territory, which provided the slogan "Fifty-four forty or fight," and the annexation of Texas were the main subjects of the campaign. It was in this campaign also that America's first "dark horse" presidential candidate was presented to the voters. He was James K. Polk, who won the nomination after the favorite, Martin Van Buren, failed to get the required two-thirds majority at the Democratic convention. Henry Clay, the candidate for the Whigs, did

not advocate war to get Texas into the Union, and this cost him the presidency. The Democratic candidate Polk and his vice-presidential running mate, George M. Dallas, won the election with 1,339,368 popular and 170 electoral votes as against Clay and his running mate Theodore Frelinghuysen with 1,300,687 popular and 105 electoral votes. (James Truslow Adams, *The March of Democracy*, pp. 359–62; Samuel Eliot Morrison, *The Oxford History of the American People*, pp. 556–57; Donald E. Cooke, *Atlas of the Presidents*, pp. 32–33; Seymour Kurtz, *The New York Times Encyclopedic Almanac 1970*, p. 148.)

4. This was known as the Chalybeate Springs, and was located several miles northwest of Jeffersonville. (Letter from Samuel W. Thomas, Director of Archives and Records Service, Jefferson County, Kentucky, to Ernst A. Stadler, July 8,1970.)

Charlestown and Jeffersonville in Indiana

1. According to Samuel W. Thomas, this was Gustavus Holland, M. D. (Letter to Ernst A. Stadler, July 8, 1970.)

2. The Oakland Course was opened in 1832, four years after Louisville was incorporated, at Seventh and Magnolia when that area "was considered quite a distance from the city," which then extended only slightly beyond Chestnut Street. (Paul S. Harris, *Fourteen Seasons of Art Accessions in Kentucky*, description of acquisition number 57.)

3. Samuel W. Thomas remarks: "I do not know what they were so happy about; Clay, the Whig, would lose the election, carrying Kentucky by about 9,000 votes and Jefferson County (Louisville) by only 50." (Letter to Ernst A. Stadler, July 8, 1970.)

4. Spelled Talles in the original German edition. George M. Dallas (1792–1864), lawyer, statesman, diplomat, United States senator, vice-president 1845–49. (Joseph G. E. Hopkins, *Concise Dictionary of American Biography*, p. 212.)

5. This date is incorrect. Koch obviously meant Sunday the thirteenth, as this entry is immediately followed by one for Monday the fourteenth.

Madison, Dupont, and Jeffersonville in Indiana

1. Koch's friend was David G. Bright (1775–1851), whose family came from Germany, where their name had been Brecht. The son to whom the letter was addressed was Jesse David Bright (1812–75), a lawyer and politician, from 1843 to 1845 Lieutenant Governor of Indiana. Jesse Bright was elected as a Democrat to the United States Senate in 1845, reelected in 1850 and 1856, and served from March 4, 1845, to February 5, 1862, when he was expelled for treason for having written a letter to Jefferson Davis in which he recognized him as "President of the Confederate States." He was an unsuccessful candidate for election in 1863 to the United States Senate to fill the vacancy caused by his expulsion. (M. S. B. Churchman, *Families of Steele and Bright*, pp. 4–9; James L. Harrison, *Biographical Directory of the American Congress 1774–1949*, p. 890.)

160

Notes

2. Spelled Wheickhofe in the original German edition. Records of the United States Census for Indiana, 1850, show a James Wycoff, 36 years old and a cooper by profession, in Jefferson County. (Letter from Miss Caroline Dunn, Librarian, William Henry Smith Memorial Library, Indiana Historical Society, Indianapolis, Indiana, to Ernst A. Stadler, June 4, 1970.)

3. Records of the United States Census for Indiana, 1830, show a James Wycoff, 40–50 years old, in Montgomery Township, Jennings County, the section through which Bear Creek flows. (Letter from Miss Caroline Dunn to Ernst A. Stadler, June 4, 1970.)

4. Although the date of the *Lucy Walker* explosion is often given as October 25, 1844, Koch records the date as October 23. Samuel W. Thomas writes: "Koch's date of October 23 is confirmed by the *Kentucky Gazette*, Vol. 58, No. 11, 26 October 1844, p. 2, Col. 5, which reported that on Wednesday last (23rd) the Lucy Walker's three boilers blew because of low water about 4 or 5 miles below New Albany, killing 60–70 persons." (Letter of Samuel W. Thomas to Ernst A. Stadler, July 8, 1970.) A more detailed description of the tragedy follows: "The principal force of the explosion took an upward direction; and the consequence was that all that part of the boat situated above the boilers was blown into thousands of pieces. The U.S. snag-boat Gophar, Capt. L. B. Dunham, was about two hundred yards distant at the time of the explosion. Capt. Dunham was immediately on the spot, rescuing those who had been thrown into the water, and affording all other assistance in his power. Having been a spectator of the scene, with all its horrors, this gentleman has furnished a narrative, to which we are indebted for many of the facts related in this article. He states that such was the force of the explosion, that, although the Lucy Walker was in the middle of the river, many fragments of wood and iron were thrown on shore. At the moment of the accident, the air appeared to be filled with human beings, with dissevered limbs and other fragments of human bodies. One man was blown to the height of fifty yards, as the narrator judges, and fell with such force as to pass entirely through the deck. Another was cut in two by a piece of the boiler. Many other incidents, equally distressing and horrifying, are related. Before Capt. Dunham could reach the spot where the wreck lay, he saw many persons who had been blown overboard perish in the water. But it was his good fortune to save the lives of a large number, by throwing them boards and ropes, and pulling them on board with boat-hooks. Immediately after the explosion, the ladies' cabin took fire and burned with great rapidity, but before it was consumed, the steamer sunk in twelve feet water. Thus the whole tragedy was completed within a few minutes." (James T. Lloyd, *Lloyd's Steamboat Directory, and Disasters on the Western Waters*, pp. 142–45.)

From St. Louis in Missouri to Bloomington in the Iowa Territory and back to St. Louis

1. In the German edition this section is erroneously numbered as chapter 11, by mistake of the printer. In the errata at the end of the book the mistake was supposed to be corrected, but the same error occurred again, and chapter 10 was again omitted.

2. This should be Sunday, November 10. Monday, November 11, is correct.

3. Here Koch again refers to the *Lucy Walker* disaster. According to Lloyd, "The books of the boat were destroyed; of course it will ever be impossible to ascertain all the names or the number of those who perished. There were at least fifty or sixty persons killed or missing, and fifteen or twenty wounded, some of them very seriously." (James T. Lloyd, *Lloyd's Steamboat Directory, and Disasters on the Western Waters*, p. 143.)

4. Spelled Patugua in the original German edition.

5. The first try at establishing a settlement at Cairo in 1818 ended in failure. A second attempt was made in 1837 when the Illinois State Legislature incorporated the Cairo City and Canal Company. The town and the bonds, issued by the Cairo City and Canal Company, were highly touted in England, where many purchased the bonds. In 1840, this scheme also failed, and the population of Cairo dwindled from about a thousand to less than one hundred. Charles Dickens visited Cairo in 1842. In his *Martin Chuzzlewit* he described Cairo, disguised as the town of Eden, as a hideous swamp, among other most uncomplimentary remarks. From Dickens's account it could be conjectured that he was one of the purchasers of the worthless Cairo City and Canal Company bonds, but there is no proof of this. (Harold L. Hitchens, *Illinois, A Descriptive and Historical Guide*, p. 172; Charles Dickens, *Martin Chuzzlewit*, pp. 174–95, 261–78.)

6. Freight barge used on the Elbe River, which rises in the mountains between Bohemia and Silesia, flows through Czechoslovakia and Germany, and enters the North Sea at Cuxhaven.

7. Keokuk (c. 1780–1848) was a Sauk leader of the Fox clan, and not a Sioux. While the Sauk and the Fox lived in Iowa country with the Sioux, they were hereditary foes, and Keokuk distinguished himself in battle against the Sioux. Keokuk was a friend of the United States government, and refused to join Black Hawk, Sauk war chief, when the latter engaged in the ill-fated Black Hawk War of 1832. After Black Hawk's defeat the United States government recognized Keokuk as the leader of the Sauk and the Fox. When in 1832 they had to cede their land to the United States, Keokuk and his followers were reserved 400 square miles on both sides of the Iowa River including Keokuk's village. In 1845 they had to give up this reservation also, and Keokuk went with his Sauk and Fox to Kansas, where he died in 1848. (Donald Jackson, ed., *Black Hawk: An Autobiography*, p. 188; Joseph G. E. Hopkins, *Concise Dictionary of American Biography*, Vol. 10, p. 350.)

8. Spelled Perlinton in the original German edition.

9. Now Muscatine. In 1833 George Davenport sent Russell Farnham to set up a store in the area; in 1836 the site of the town was surveyed, and the town was called Bloomington after the Indiana home of one of the early settlers, John Vanetta. In 1849 the name was changed to Muscatine. (Raymond Kresensky, *Iowa, A Guide to the Hawkeye State*, p. 290.)

10. Nauvoo was to the right on the trip upriver; actually it is on the left bank of the Mississippi.

11. Spelled Prott in the original German edition. Dale L. Morgan, historian, stated: "Mr. Prott was evidently Parley P. Pratt. He had been in Liverpool in the early 1840s, and

162

Notes

---

there and at Manchester published *The Latter-Day Saints' Millenial Star.* His brother, my great-grandfather Orson Pratt, did not take charge of the British mission until 1848." (Conversation between Dale L. Morgan and Ernst A. Stadler, 1969.)

12. A check of the available records of St. Louis Protestant churches then in existence has failed to turn up any membership or baptismal listings of the Koch family.

Sulphur-Springs and Herculaneum in Missouri

1. This is now the Fred S. Anheuser estate, Kimmswick, Missouri, Windsor Harbor area. (Letter from Frank Magre, a trustee of the Missouri Archaeological Society and an authority on Indian petroglyphs, Crystal City, Missouri, to Ernst A. Stadler, September 7, 1970.)

2. Spelled Harington in the original German edition. In Pevely, in Joachim township, the very earliest settler was Bartholomew Herrington, who located there in the fall of 1799. His son, Joshua Herrington, was born in Illinois in 1800 but was brought to Jefferson County a month or so after his birth and continued to reside there. (Mary Joan Boyer, *Jefferson County, Missouri, in Story and Pictures,* pp. 60–61.)

3. This should be Friday, the 29th of November.

4. These petroglyphs are still to be seen on rock ledges at the confluence of the Joachim Creek with the Mississippi River, right side, in an area to which access is owned by the St. Joseph Lead Company of Herculaneum. Herculaneum was moved to the left bank, so the house with the footprint on the rock in the chimney would no longer be there. (Letter from Frank Magre to Ernst A. Stadler, September 7, 1970.)

5. Herculaneum, founded by Moses Austin in 1809, is said to have been the first settlement in Jefferson County to grow into a town. Moses Austin applied American enterprise to mining and smelting of lead at Mine au Breton (Potosi) and gave Herculaneum its start as a landing on the Mississippi River. He brought supplies there for his Washington County operations and from Herculaneum he shipped lead. Making of shot became an early industry, along with distilling whiskey, and St. Louis provided a market for farm products of the vicinity bringing about a rapid growth for Herculaneum. Nothing remains of the original town, which was swept away by the river. In 1896 the St. Joseph Lead Company opened a large lead and zinc smelter at Herculaneum, which reestablished the historic settlement. (Mary Joan Boyer, *Jefferson County, Missouri, in Story and Pictures,* p. 104.)

6. No account could be found of this incident in local newspapers.

7. Spelled Helsborough in the original German edition.

8. Spelled Mamouth in the original German edition. The Mammoth mines were six miles southwest of DeSoto, Missouri, at the confluence of Mammoth Creek and Big River. (Henry C. Thompson, *Our Lead Belt Heritage* [Flat River, Mo., 1955], p. 83, quoted in a letter from Frank Magre to Ernst A. Stadler, September 7, 1970.)

9. These footprints were undoubtedly those at the present Washington State Park near DeSoto. At that time they were on a part of the farm belonging to George Higginbotham. There are many footprints still at this site. (Letter from Frank Magre to Ernst A. Stadler, September 7, 1970.)

163

10. According to Frank Magre, these footprints were cut from the many still at the confluence of the Rock Creek and the Mississippi River on the present Fred S. Anheuser estate at Kimmswick. (Letter from Frank Magre to Ernst A. Stadler, September 7, 1970.)

11. Spelled Water in the original German edition. Captain George Waters, a graduate of the United States Military Academy at West Point, was assigned to Jefferson Barracks, resigned from the army in 1838, and moved to what was eventually to be known as Windsor Harbor, at Kimmswick. In 1840–41 he built a house near the Mississippi almost entirely of oak lumber put together with wooden pegs. There were eight or ten rooms, and every room had a fireplace large enough to hold six-foot logs. The old house eventually passed out of the hands of the descendants of Captain Waters, and became the property of Fred Anheuser and, still later, of his son Fred. In 1951 the old landmark, said to be one of the oldest houses in in the Kimmswick-Windsor locality, was torn down, and some of the salvaged lumber was used in the residence of the Anheusers nearby. (Mary Joan Boyer, *Jefferson County, Missouri, in Story and Pictures*, pp. 81–83.)

12. There was a petroglyph site at a point two miles south of Herculaneum at a place called Plattin Rock at the edge of the Mississippi. It is now filled over, and the elevated land is used as a sand recovery plant by the Fred Weber Construction Company. (Letter from Frank Magre to Ernst A. Stadler, September 7, 1970.)

Golconda in Illinois; Smithland in Kentucky; Natchez, New Orleans in Louisiana

1. This occurred near Grand Tower. (Albert C. Koch, *Description of the Hydrarchos Harlani*, p. 6.) Grand Tower is an island in the Mississippi River, about 100 miles south of St. Louis opposite the present-day town of Grand Tower, Illinois. The Tower Rock, for which the island is named, rises a sheer 60 feet from the waters near the Missouri shore, and its top is about an acre in extent. The island, which is under federal control, is sometimes referred to as "the smallest national park." (Henry Lewis, *The Valley of the Mississippi Illustrated*, p. 338; Katharine and Will Griffith, *Spotlight on Egypt*, p. 16.)

2. Here Koch alludes to Christmas in Germany, where the twenty-fifth and the twenty-sixth of December are official holidays. The twenty-fifth is referrred to as the first, and the twenty-sixth as the second, Christmas holiday.

3. Golconda, principal town and county seat of Pope County, close to the bank of the Ohio River. Originally called Sarahville, Golconda was not a vigorous village, and by 1836 had but three stores, one grocery, two taverns, and about twenty dwellings with about one hundred people. (George Washington Smith, *History of Southern Illinois*, p. 516.)

4. Records in the Golconda (Pope County) courthouse show a Hiram J. Graham and Albert Koch engaged in the sale of land. The residence of Graham is given as Pope County, and later as Fulton County, Illinois. (Deed Record C, 1845, pp. 285, 287; Deed Record D, 1848, p. 312.)

5. Smithland, Kentucky, once a promising town at the confluence of the Ohio and Cumberland rivers. The decline of Smithland began with the fading of the steamboat era and was sealed by the building of the railroads, which bypassed the town. (U. R. Bell, *Kentucky, A Guide to the Bluegrass State*, pp. 411–12.)

Claiborne in Alabama

1. Claiborne, Alabama, a town on the site of Fort Claiborne, which was built by General F. L. Claiborne in 1813 as a supply base for his invasion of the Creek country. (Howell Tatum, "First Alabama River Survey 1814," *Alabama Historical Quarterly* 19, no. 2, Summer 1957, p. 218.)
2. William Thomas Hamilton, a Presbyterian, born in England, was pastor of the Government Street Presbyterian Church in Mobile from 1834 to 1854. (Letter from Milo B. Howard Jr., Director, State of Alabama Department of Archives and History, Montgomery, Alabama, to Ernst A. Stadler, July 24, 1970.)
3. Timothy Abbot Conrad, 1803–57, American geologist, paleontologist, and naturalist. (Norma Olin Ireland, *Index to Scientists of the World from Ancient to Modern Times*, p. 140.)

Macon, Clarksville, Coffeeville, Washington-Old-Courthouse, St. Stephens, Mobile in Alabama

1. Probably one of three brothers, Joseph, William, and Elijah Chapman, who had come from South Carolina and were engaged in farming near Grove Hill in Clarke County. (Letter from Milo B. Howard, Jr., to Ernst A. Stadler, July 24, 1970.)
2. Spelled Washborne in the original German edition. This was probably George R. Washburn, who married Lucretia Lide in Clarke County, April 1, 1840. The 1850 Clarke County census lists a G. R. Washburn family. (Letter from Milo B. Howard, Jr., to Ernst A. Stadler, July 24, 1970.)
3. Spelled Paket in the original German edition. Sir Charles Lyell mentions that he was fortunate enough to meet Mr. William Pickett, "who had most actively aided Mr. Koch in digging up the skeleton of the fossil whale, or zeuglodon, near Clarkesville." William Pickett was commissioned justice of the peace May 27, 1843, and again March 14, 1844. (Sir Charles Lyell, *A Second Visit to the United States of North America*, Vol. 2 , p. 59; *Alabama Secretary of State's Commission Register*, 1832–44, Vol. 2, p. 68, and ibid., 1844–67, Vol. 3, p. 116; letter from Milo B. Howard, Jr., to Ernst A. Stadler, July 24, 1970.)
4. Spelled Tattilliby in the original German edition. Mr. Howard writes: "Tattilaba Creek is a tributary to Jackson Creek in Clarke County. The spelling Tattilaba is from William A. Read's *Indian Place-Names in Alabama*. As is the case with Tombigbee (the present spelling of the river), all Indian names were spelled phonically by the French first and then by the English. Add to that combination a modicum of illiteracy and various kinds of Southern drawls, and the result is chaotic. You may choose almost

arbitrarily the spelling that you prefer." (Letter from Milo B. Howard, Jr., to Ernst A. Stadler, July 24, 1970.)

5. Again quoting Mr. Howard, "On the LaTourette map of Alabama, 1838, Washington Court House is located in Range 3 West, Township 8 North, Section 9, St. Stephens meridian, on the road from St. Stephens northwest. Prairie Bluff is in Range 7 East, Township 14 North, Section 31, St. Stephens meridian, on the west side of the Alabama River in Wilcox County. Washington-Old-Courthouse is a quaint way of saying Old Washington Courthouse. The Court House had been moved from Washington Courthouse, hence the 'old.'" (Letter from Milo B. Howard, Jr., to Ernst A. Stadler, July 24, 1970.)

6. Spelled Krie in the original German edition. Later in this chapter Koch spells correctly the name of Judge John G. Creagh, 1787–1839, lawyer, farmer, legislator, and probate judge. In 1834 and '35 Judge Creagh had found *Basilosaurus* bones on his plantation and had sent them to Timothy Abbot Conrad; the bones were subsequently studied by Dr. Richard Harlan. In 1842 Samuel B. Buckley, botanist and field naturalist, excavated a vertebral column of *Basilosaurus cetoides*, which extended to a length of 65 feet, on Judge Creagh's plantation. (Letter from Milo B. Howard, Jr., to Ernst A. Stadler, July 24, 1970; Joseph G. E. Hopkins, *Concise Dictionary of American Biography*, p. 120; Remington Kellogg, *A Review of the Archaeoceti*, p. 16.)

7. Another version of this story is that one Israel Slade had found *Zeuglodon* remains on Judge Creagh's plantation in the spring of 1845, and that Koch had bribed Slade's men to let him have the bones. (Remington Kellogg, *A Review of the Archaeoceti*, p. 17.)

8. Spelled Tombeckbe in the original German edition.

9. Daniel M. Williamson, a farmer living near Fail, Alabama, reported to Charles Schuchert, a geologist and paleontologist who lived from 1858 to 1942, that he had been the boy who directed Koch to places where *Zeuglodon* bones were to be found. (Norma Olin Ireland, *Index to Scientists of the World from Ancient to Modern Times*, p. 544; Remington Kellogg, *A Review of the Archaeoceti*, p. 17.)

10. Sintabouge, a small river which empties into the Tombigbee near the northeast corner of Washington County. (*Alabama State Gazetteer and Business Directory*, Vol. 3, map.)

11. Spelled Chicatahay in the original German edition.

From Mobile in Alabama to St. Louis in Missouri and from there to New York

1. Koch has confused several dates in this chapter. This one was either Tuesday, June 17, or Wednesday, June 18.

2. Friday, June 27, or Saturday, June 28.

3. The great fire in Pittsburgh occurred on April 10, 1845, raging over thirty-six acres and destroying twenty blocks of buildings. The property loss amounted to more than $5,000,000 and made two thousand persons homeless, but only two lives were lost. (C. C. Lesley, *Pennsylvania, A Guide to the Keystone State*, p. 299.)

4. Samuel George Morton, 1799–1851, physician and naturalist who described the fossils brought back by the Lewis and Clark expedition. (Joseph G. E. Hopkins, *Concise Dictionary of American Biography*, p. 707.)

5. One brief published account of the wreck and salvage operation follows: *"Key West June 19*—The decree in the case of the cargo and materials of the wrecked ship *Newark* gives to the salvor 40% upon the estimated value of the property saved, after first deducting costs of suit, wharfage, storage and charge for labor. This estimated value is $12,173.07 The cost and charges are not yet ascertained. Among the cargo saved are 28 boxes of fossil bones marked D. A. Rock [Dr. A. Koch?] upon which no salvage has been decreed. Supposing these fossils may not have been shipped for profit or speculation, but for the benefit of science, the wreckers have resolved to ship them to Messrs. E. D. Hurlbut and Co. of New York, with instructions to deliver them to the original consignee, free of any charge for salvage, if it shall appear that they are in fact only designed for scientific purposes." (*New-York Commercial Advertiser*, July 1, 1845.)

6. Friday, July 4, or Saturday, July 5.

7. This conflagration is wryly described in an 1880 account: "On the 19th of July, 1845, another great fire, second only in its ravages to that of 1835, broke out in New street in the vicinity of Wall, and burned in a southerly direction to Stone street, laying waste the entire district between Broadway and the eastern side of Broad street, and consuming several million dollars' worth of property. The explosion of a saltpetre warehouse in Broad street during this conflagration, gave rise to the vexed question, 'Will saltpetre explode?' which furnished food for some research and much merriment to the savans of the day." (Mary L. Booth, *History of the City of New York*, p. 748.)

8. Tuesday, August 5, or Thursday, August 7.

Richmond, Petersburg in Virginia; Rockymount, Gaston, Enfield in North Carolina; Grove-Landing, Norfolk in Virginia; Baltimore in Maryland; Philadelphia in Pennsylvania; Albany, Utica, Trenton in New York

1. Enfield, once known as Huckleberry Swamp, is the oldest town in Halifax County. (Edwin Bjorkman, *North Carolina, A Guide to the Old North State*, p. 319.)

2. Fishing Creek borders Halifax County near Enfield. The bones found here were those of an *Ichthyosaurus*. (Ibid., p. 319.)

3. Monday, December 22, or Tuesday, December 23.

4. Spelled Whiticker in the original German edition. The name Whitaker is quite prevalent in this area. The small town of Whitakers was named for Richard and Elizabeth Carey Whitaker, the first white settlers to venture into this Tuscarora stronghold. (Edwin Bjorkman, *North Carolina, A Guide to the Old North State*, p. 319.)

5. Saturday, December 27, or Sunday, December 28.

6. The early settlement of Rocky Mount was named after the mounds near the falls of the Tar River. It was here, at Donaldson's Tavern, that Lafayette spent a night

when he was on his triumphant tour of the United States in 1825. Koch's uncomplimentary description of the place calls to mind the story told about a barkeeper who was supposed to have burned down the Falls Primitive Baptist Church after he learned of a law which forbade the sale of spirits near a place of worship. The settlement developed after the Wilmington and Weldon Railroad had laid tracks through the area and slowly merged with the village around the cotton mills mentioned by Koch. The town, incorporated in 1867, lies in two counties; Main Street is split down the middle by the county line, with the result that the citizens on one side of the street have to go to court in one county, while those living on the other side have to attend court in the other county. (Edwin Bjorkman, *North Carolina, A Guide to the Old North State*, pp. 334–35.)

7. In 1646 Fort Henry was built on the spot where Petersburg now lies, on the south bank of the Appomattox River. It was named after Peter Jones, an early Indian trader who had a trading post there called Peter's Point, which was later changed to Petersburg. The town is large and flourishing, and lies on the great southern railroad, twenty-two miles south of Richmond and nine miles southwest of City Point. (Henry Howe, *Historical Collections of Virginia*, p. 243.)

8. City Point, on the James River at the junction of the Appomattox, was once a not unimportant, though small, village at the outport of Richmond and Petersburg. Its wharves could accommodate ships of the largest classes, and there was considerable river traffic frequenting the place. City Point was considered a much better site for a commercial town than Richmond, and it could have become the capital of Virginia, but its owner, "a Dutchman," refused to sell it for any price. (Ibid., p. 440.)

9. Captain John Smith landed with the colonists of the Virginia Company of London at Jamestown in 1607. His adventure of being captured and threatened with death by the Indians, and his dramatic liberation through the intervention of Pocahontas, he told in his *Generall Historie*. Although the story of his rescue has assumed legendary proportions, there is no reason to doubt its essential truth. Jamestown, the first seat of a legislative assembly in America, fell into decay after the removal of the capital to Williamsburg in 1699. (Joseph G. E. Hopkins, *Concise Dictionary of American Biography*, p. 973; Wayne Andrews, *Concise Dictionary of American History*, pp. 496–97.)

10. This splendid edifice had evolved from an earlier institution known simply as Peale's Museum. Charles Willson Peale, 1741–1827, popular portrait painter of his time, was authorized in 1802 to use rooms in Independence Hall as a museum. In 1821 Peale's Museum was incorporated into the Philadelphia Museum Company. On July 4, 1838, the Philadelphia Museum Building, at the northeast corner of Ninth and Sansom Streets, was opened. The building was two stories high with lofty ceilings, a length of 238 feet, and a width of 70 feet. The museum contained a large collection of objects of natural history and a portrait gallery with several hundred paintings, many by the elder Peale and some by his sons. On the second story, in the center of a long room, was displayed the skeleton of "the great mammoth." In 1799, the elder Peale had obtained a great number of bones of extraordinary size from a marl pit near Newburgh, New York, on the Hudson River. With the help of his sons

Rembrandt and Raphael Peale one skeleton was assembled, and, because there were enough bones left over, the brothers constructed another one, which they took to England in 1802. Unsuccessful in their efforts to sell it there, they brought it back one year later. The museum, meanwhile, fell on hard times, and in 1842 and '43 the lower room was opened for public use, and balls, public meetings, and concerts were held there. After about six years the Philadelphia Museum was forced to sell its collection of natural history objects, the largest in the country. Now the whole building was used for such events as the annual floral show of the Horticultural Society and the annual exhibition of the Franklin Institute of stoves, grates, and machinery. In 1847 the Whigs held their national convention at the museum, and in 1854 the building burned to the ground. (J. Thomas Scharf and Thompson Westcott, *History of Philadelphia 1609–1884*, Vol. 2, pp. 947–50.)

11. A telegraph line was completed from New York to Philadelphia in 1845, to Boston in 1846, and to Albany in 1847. (Moses King, *King's Handbook of New York City*, p. 41.)

12. Saturday, May 2, or Sunday, May 3.

Index

Academy of Science of St. Louis: Koch elected to, xxxi

Alabama: description of, 96–99; Koch discovers *Zeuglodon* remains in, xxviii–xxix, 96–107, 108; Koch told of *Zygodon* remains in, 32; places visited in, xxviii–xxix

Alabama River: description of, 109; explored by Koch, 108; low water level of, 92

Albany, N.Y.: Koch visits, 131–32; mentioned, 135, 136

American Philosophical Society, xxvi

Amos, the Rev. Joseph (blind Indian preacher): facts about, 156; preaches on Gay Head, 19–20, 24

Anheuser, Fred S.: petroglyphs on estate of, near Kimmswick, Mo., 162, 163

Apollo Rooms, New York: Koch exhibits *Hydrarchos* in, xxix

Arkansas River, 111

Atwater, Caleb: facts about, 38, 158

Austin, Moses: facts about, 162

Babcock, George: makes geological explorations with Koch, xxxii

Baltimore, Md.: Koch disassembles *Hydrarchos* in, 131; mentioned, 122, 123, 130, 131

Baptisms: in ocean at Gay Head, 20; in Ohio River, 113; Mormon, 67

Barbecues: at Jeffersonville, Ind., 51; at Utica, Ind., 60

Barges: escort Steamboat *Uncle Toby* in low water, 63

Basilosaurus. See Hydrarchos

Bear Creek, Ind.: Koch visits, 56, 57

Benton County, Mo.: exhumation of *Missourium* by Koch in, xxiv–xxv; Koch's explorations in being reevaluated, xxxiv

Berlin: *Hydrarchos* destroyed by fire in, xxxv; *Hydrarchos* exhibited in, xxx; *Hydrarchos* purchased for Royal Anatomical Museum of, xxx

Beyer, Moritz: collaborates with Louis Koch on book, xiii

Bloomington, Iowa: facts about, 161; Koch visits, 66–67

Boonville, Mo.: *Missourium* shipped to St. Louis from, xxv

Boston, Mass.: Koch visits, 12–13, 29–30

Breslau: Koch's second *Zeuglodon* exhibited in, xxxi

Briggs, George Nixon: facts about, 157; visits Gay Head, 24

Bright, David G.: facts about, 159; writes letter of introduction, 53

Bright, Jesse: facts about, 159

Buckley, Samuel B.: excavates *Basilosaurus* remains, 165

Buffalo, N.Y.: German religious sect near, 132; Koch visits, 34

Bunker Hill Monument: description of, 13

Burlington, Iowa: Koch visits, 65

Cairo, Ill.: facts about, 62, 161; Koch visits, 62; mentioned, 111

Cambridge, Mass.: description of, 29–30

Camp Creek, Ind., 56

Canal boats: Koch travels on, 33–38, 114–15

Carus, Carl Gustav: facts about, 154; to study *Hydrarchos*, xxx

Catlin, George: paints and describes Osceola, xx

Caves: Koch explores cave near Claiborne, Ala., 91

Cemeteries: Grove Street Cemetery, New Haven, Conn., described, 11, 155